COLD

COLD

THREE WINTERS AT THE SOUTH POLE

WAYNE L. WHITE

Potomac Books

AN IMPRINT OF THE UNIVERSITY OF NEBRASKA PRESS

CONTENTS

ILLUSTRATIONS

COLD

THE GLORIOUS ICE

The Amundsen-Scott South Pole Station is a ship. A ship with a crew that sits atop nearly two miles of moving ice. The ship, the ice that holds it, and its crew are on a voyage. A voyage that will take several hundred thousand years before whatever is left of the ship along with the ice that held it will arrive at its final destination, the Southern Ocean. Its winter crew will change out annually and are on a voyage of their own. That voyage will test some of them to their limits, and all, to various degrees, will be forever changed by the experience. While their icy voyage will not yield views of distant lands, it will give some crewmembers remarkable insight into unfamiliar places. Those places will include who they really are as people and how their being impacts others. The crew will reside in their ship in nearly complete physical isolation from the world they knew. While inside, it will protect them from unearthly cold that few on the planet will ever experience. They will lose the sun and experience darkness that will last for nearly half the year they are there. The ship and its crew will supply the light. I had the great honor of being the captain of that ship for three Antarctic winters.

At the South Pole, at around minus 80 Fahrenheit, a strong exhalation will result in a strange whooshing noise from a very visible cloud coming from your mouth. At first, this noise surprises you as you have breathed many thousands of times and never heard anything like it. The sound's origin is the fact that the water vapor from your breath is freezing incredibly rapidly and hurtling through the air. While out walking across the ice, you are startled by a loud cracking noise that sounds as if things are collapsing and you will fall into oblivion. The cause of the noise is a small horizontal void of air collapsing within the ice. Much to your relief, you do not disappear below the ice and walk on. When you are walking into the wind and not wearing goggles, you start to develop icicles dangling from your eyebrows and eyelashes. If you are out long enough, these may need periodic removal as vision can be obscured. From time to time, especially when facing into the wind, your eyelids may temporarily freeze shut. Squinting hard, shutting your eyes tighter will break the tiny ice seal, and much to your relief your eyes can be opened. You can see again. I experienced those things while at the South Pole and became used to them. I have heard of another phenomenon that occurs when ice crystals are in the air in a certain configuration. When this happens, it increases acoustics to the point where you can hear normal conversations from long distances. I never experienced that. Outside, I was always alone.

The ice and beauty of Antarctica did not call out to me. I never liked the cold, but during my time at the South Pole, with much exposure to unimaginable cold, I would eventually love it. I would love it while still never liking it. Given any thought, cold is the opposite of life. When things die, they cool. Life for humans is about warmth, and it is abnormal to seek the cold. I learned to love and to embrace the cold, but it took time and the ability to recognize what it gave me. In the Midwest, where I grew up, there were very distinct seasons, and I well remember after the muggy heat of summer the euphoria of a chilly autumn with sweaters and jackets. Then came the onset of winter and most years the wonders of a white Christmas. After New Year's, the novelty of the snow, ice, and cold wore off and months of bleakness followed. Most birds, other animals, and humans took refuge to await the spring, and the white and icy world, to me, was cold and sterile.

Initially, my life took me to warmer places, starting with leav-

ing the Midwest for sunny California and the brutal experience of
USMC boot camp. I never returned to the Midwest and called California home for several years before leaving for the warm tropical
climate of Diego Garcia British Indian Ocean Territory, where I
spent over five years living and working. This led to further foreign
overseas or U.S. remote site adventures and another fifteen years in
places such as Midway Atoll, Iraq, Kuwait, Wake Island, Afghanistan, Saudi Arabia, and finally, Ascension Island South Atlantic.
The only anomaly in these mostly warm weather places where I
lived and worked was the nearly four years I spent on rugged Shemya Island, Alaska, near the end of the Aleutian chain. While certainly not tropical, Shemya did not have the frigid conditions of
Alaska's interior. It was about as cold as I wanted it to be.

As a boy I was fascinated by geographical exploration and
dreamed of travels to exotic places such as jungles, deserts, and
tropical islands. I did not dream of cold places. As an adult, I was
able to travel the globe and even launch my own minor expeditions
in tropical places such as New Guinea, the Amazon, and Africa.
Historic expeditions of the past had always intrigued me. I was
drawn to anything that had to do with the search for the source
of the Nile, the crossing of equatorial Africa, and expeditions into
remote parts of the Amazon basin. Physical struggle itself attracted
me, along with the great dangers that had been faced by men to
penetrate the unknown. I always had a strong sense of adventure
and a love of history, to which I added a physical fitness regimen
and an outlook that ensured there was no place on the planet that
I could not go. During my own travels and minor expeditions,
I began collecting ethnic- and exploration-related artifacts, and
over the years I amassed quite a collection in my home in coastal
Texas. While doing this, I became aware of another great exploration struggle, and that involved ice.

The great race to the South Pole that occurred in 1910–12 very
much intrigued me. In that epic struggle, the Norwegian Roald
Amundsen, his tough men, and their tough and edible dogs had

won the race to the South Pole, but in that race, Captain Robert Falcon Scott and his men had become immortal. In order to become immortal, Captain Scott and his polar party had to perish on their return from the South Pole. This was after the realization that Amundsen had arrived there first. Being first to the South Pole was a major goal at that time in history and no one wanted to be second. During what must have been an incredibly depressing march back to their Cape Evans hut, which was over eight hundred miles away, Scott and his men died due to lack of food, fuel, and the colder-than-normal temperatures that occurred during that year. Before dying with his two remaining comrades in his frozen tent, Captain Scott wrote extremely poignant letters to various wives, now to be widows; a mother soon to be without her son; his financial backers; and the British public. This epic drama that occurred in what has been termed "the heroic age" of Antarctic exploration was my real attraction to the cold.

After Captain Scott departed the South Pole on January 19, 1912, no one saw the need to return to a place that he had described in his diary with the words "great God this is an awful place!" He then perished with his polar party on the return journey, and man did not again visit the South Pole until October of 1956. On that date, the U.S. Navy landed a R4D-5 Skytrain aircraft that spent forty-nine minutes on the ice. This opened the South Pole to a new type of explorer quite different from those that visited during the earlier heroic age of exploration. The new explorers were also heroic but had modern technology and a new scientific mission. Now the United States National Science Foundation, NSF, manages the United States Antarctic Program, USAP, which was founded in 1959. The USAP is responsible for the upkeep of three Antarctic stations, numerous field camps and ships. It is responsible for coordinating all the science conducted by several agencies and educational institutions. It also administers the Antarctic Support Contract, ASC, headquartered in Denver, Colorado, as the USAP is heavily dependent on contractors to conduct its mission. This was my route to the ice.

My gateway to the ice came by way of the tropics. While serving as the station manager on a United States Air Force contract on Ascension Island, South Atlantic, one day I noted an employment advertisement seeking a winter site manager, WSM, at the South Pole. Several years earlier, I had applied for that position and been selected as an alternate. I remembered the past interview at the Denver ASC office and how I had made an impassioned attempt to be considered for the primary role. Unfortunately, I was up against someone who had already been there and could not compete against such experience. I was then offered the alternate role, which I accepted, but I did not deploy as the primary candidate did. Going through that process gave me a solid understanding of the program and how, with my education and experience, I thought I could fit. The WSM serves as the station's leader during winter, which at the South Pole is most of the year. At the South Pole, summer runs from the beginning of November to mid-February and the winter is everything else.

Thinking back, I cannot remember why I had even seen that advertisement. I had no reason to look for another position. I had wanted the Ascension Island position very badly and waited over two years for the government to award the contract. During the bidding process, I had been listed in Alaska Native Corporation's proposal, stating that in the event they won the new contract, I would be the island's station manager. I had previously worked for that company on Midway Atoll, Shemya Island, Aleutians, and Wake Island. Over a two-year period, I would periodically be informed of the contract's award status. There had been several setbacks as contract extensions for the existing contractor were awarded. As the process dragged on, I was viewing the opportunity as quite a longshot and was then somewhat shocked and immensely thrilled when we won it. The attractiveness of the Ascension Island contract was, for me, immense. It had everything I wanted in a project: it qualified for the U.S. federal tax exemption, as it was an overseas assignment; travel there from my home in Texas was relatively easy, flying out of Patrick Air Force Base in Florida on a contractor flight; and most importantly, the island was beautiful and had an amazing history. After other places I had been in the world, including two wars, it was truly paradise. Now, I had that great job, had lived and worked there for over eight months, and loved the place. There was absolutely no reason for me to leave—and yet, I did.

One of the things I told myself to justify my applying for the South Pole WSM position was that my application would probably go nowhere as someone had probably already been informally selected, as had happened last time. I had nothing to lose and probably was not going anywhere. Thinking that, I was somewhat surprised with an almost immediate contact by people from the Denver ASC office who asked a few questions and then very rapidly set up a face-to-face interview. At that point, I realized that this was serious and that I might have to commit. Thinking of departing the beautiful island I had waited so long for was slightly sickening. Even so, I was very aware this could be an adventure that would not come my way again. My career up to that point had included several changes where I left great island assignments to experience new things, such as the wars in Iraq and Afghanistan. I had been bold and seldom sentimental in making these changes. So far, that had worked for me.

One of the things that motivated me to make the radical change from known paradise to the unknowns of the ice was my collection in my house in Rockport Texas. The collection started with ethnic artifacts brought back from my first expeditions to New Guinea, the Amazon, and Africa. Through the years, with all my travels, it had grown to include other areas of the world and exploration-related items that I found interesting. By this time, I owned several items from the estate of the great African explorer Henry Morton Stanley, including furniture, weapons, and his personal ring commemorating his 1874–77 Anglo-American expedition across Africa. The collection also featured a small piece of an African tree from central Africa, beneath which the great missionary explorer Dr. David

Livingstone had his heart buried. The house was packed with such things, and they were on open display, hung on walls, or residing in glass cabinets. A full-size bronze museum bust of my guide in life, Richard Francis Burton, resided in a prominent place looking down upon it all. I filled my house with these treasures to the point that it was a minor spectacle to most people upon entry. Most of the items I had an affinity for or connection with as I had traveled to the locations, and with that had a better understanding and appreciation of the objects. In the back of the house, I had a "polar room" dedicated to the great deeds of polar explorers in both the north and the south. While I had some minor experience in Alaska, I had never been to Antarctica. Over time that began to bother me more and more, and at one point, thinking I would never go, I considered removing that part of the collection to make room for items from places I had been. Fortunately, I did not and now could fix that.

While at Ascension Island there were many things to do, from exploring the island's early history to hiking to the Green Mountain summit. Charles Darwin had been there and wrote about its biology. In the 1800s it had served as a place to convalesce for sailors who were exposed to tropical diseases in Africa. Many times, they died and were buried there. One of the activities I always enjoyed was reading, and at Ascension I read Herman Melville's classic *Moby Dick*. That novel and its message about an obsessed and isolated sea captain would loom large in my future.

I was at my eight-month point on the Ascension Island project and was due a home leave of a couple of weeks. My plan was to fly to Denver first and do the South Pole interview. I told no one on the Ascension project that I was doing this and planned to take it one step at a time. This would start with completing the face-to-face interview in Denver and a hiring decision by the ASC. If not selected for the position, no one would ever need to know. If selected, I would finish my leave, return to the island, and give a one-month notice. It was an unsettling departure from that beautiful island, as my future, which had seemed so solid, now might face a major change. My flight from Ascension Island to Patrick Air Force Base ran late and led to an extremely tight connection with my commercial flight to Houston departing from Melbourne, Florida. I barely made that flight, and my day was far from over. After a late arrival in Houston, I had a few hours of fitful sleep in an airport hotel and an early morning flight to Denver.

Exhausted upon arrival in Denver, I checked into a hotel that I barely remember. The next morning was the interview, and although exhausted, I did my best to represent what I was. By this time in life, I had come to the point that I would not deceive or hold back my being from any prospective employer. Interviews can be episodes in deception with candidates who strongly desire a position and try to present themselves as the perfect fit. I was way past that point. I wanted a prospective employer to fully understand and consider what I was and how I would or would not be what they were seeking. I had worked over twenty years as a U.S. Defense Department contractor at mostly remote locations where the relationship among a leader and employees is far different from a standard U.S. job. I had been involved with tough projects, adversarial relationships with contract customers, had people die, and had thousands of incredible experiences far from the U.S. eight-to-five norm. This along with several other extreme undertakings in life had made me what I was and there was no need to hide it. While I always respected my company superiors, I found I worked best when they were thousands of miles away and provided minimal to no direction. I worked alone. I did not hold anything back during the interview and had prepared myself for a rejection, as by this time my fit in most of civilized society was a bit of a stretch. If there was one thing I did want to come across in my interview, it was that my employees and what they needed came first. I was not soft with employees and was ruthless with malignant personnel who could infect others, but I found most of the people who had worked for me were good people and deserved my best. I do not remember a lot of the Denver office or the interview except for the question about why I wanted to do it, and my answer was "because I can't go to Mars." I will always remember the kindness I received during my interview and the short time I was in that office. I was then given an incredibly priceless gift, my selection for the position.

I returned to Ascension Island, resigned my station manager position, and prepared for my departure. Interestingly, the management at the Alaska Native Corporation I was working for quite understood my early departure from the project and the significance of the South Pole position. Leaving beautiful Ascension Island and the wonderful people I had met while there was not easy. Making it harder was leaving around the nine-month point in my contract and not fulfilling a year overseas; I would lose what would have been a sizeable federal tax refund. These were both strong reasons to stay, but not strong enough for me. I had been on several fascinating islands over the years and had been on Ascension long enough to explore most of it and gain a good understanding of its wonderful people and history. I had spent many years living, working, and exploring the tropics and felt it a natural environment for me. That was well and good, but I felt I was deficient in my experience and understanding of the ice. I wanted to more fully understand the conditions that my great heroes Amundsen, Scott, and Shackleton had faced. The polar room in my Texas home was impressive, but I had not been to the areas on display; I did not understand the ice. It was foreign to me. I would now understand the ice and what extreme cold felt like. I would also get a chance to use my leadership style in the closed, frigid environment at the South Pole.

At that point, my leadership style was a combination of three things. The first and most minor component was what I had learned during my single enlistment in the U.S. Marines many years earlier. During those years I had seen few of what I would describe as great leaders. What I saw was a system with blind obedience to rank. During my enlistment, I witnessed a few idiotic officers and, at times, thuggish and somewhat dim enlisted personal being obeyed simply due to the rank structure. For the Marines, with its mission, it worked. In a crisis, when time is of the essence, the system has its merits. I learned two valuable life lessons while in the Marines. The first was that I would never again allow myself to be in a situation where anyone or any organization had the type of life-and-death power over me that I experienced while under Marine command. The second was that I would be a Marine until the day I died and would always be extremely proud of my service. While I greatly respected the Marines as an organization and was proud of the time I served, I knew that leadership style was not what worked in the civilian world, except in the direst of situations. If I found myself in a dire situation, it was good to know I could revert to that style of leadership.

Next, I had studied the world's great explorers, focusing mostly on the nineteenth and early twentieth centuries. A trait that most of those explorers shared was that they were egomaniacs hell-bent on fame and fortune. How they were able to motivate others to assist them in that quest is where they differed. With Antarctica, my main interest was "the heroic age of exploration," where the leadership styles and results could be studied between the three greats: Roald Amundsen, Captain Scott, and Ernest Shackleton. All great men with vastly different styles of leadership. The cold, distant, and precise Amundsen. The virtuous, steady, and brave Captain Scott. The gregarious, likable, and caring Ernest Shackleton. I loved them all for what they were and felt some combination of the three would probably be a perfect leader. One challenge in using them as guides for leadership was that people have changed since their days. What motivated someone to follow a leader in the early 1900s might be quite different today.

The last and probably most important facet of leadership was what I had gained from more than twenty years working for various bosses at U.S. Defense Department assignments around the globe. I had been extremely fortunate to work for many great men, and

I had watched closely how they interacted with the various work forces and customers. I took note of what worked and what did not. I was fortunate to have so many great men who tolerated me and taught me what they did. I never forgot what they had done for me and now owed them my undying loyalty. I also owed it to them to do my best to someday do the same for people who worked for me. By the time I joined the USAP, I had made many mistakes through the years at projects I had worked on, but I had gained priceless experience. I felt I was ready, but I wanted to proceed cautiously and was not overconfident. I did not yet know the ice.

DENVER

Denver, more specifically, the town of Centennial, Colorado, is where the ASC has its headquarters. Upon arrival in Denver to begin my new position, I checked into a long-term hotel not far from the ASC office. It was far enough away that it required a van ride every day. This ride ended up being one of the high points of my Denver stay. Upon reporting to the ASC office, my education in ice commenced. The office was large, occupying two stories. It was filled with people in cubicles and small offices who worked in various parts of the massive program. Most of these people would, for me, remain nameless faces. There were several Antarctic displays in cabinets, a mannequin dressed in extreme cold weather clothing, and photos and paintings of an icy world where I had never been. I was to learn that the word *Denver* would loom large in current Antarctic terminology and could apply to all manner of things, both positive and negative. The primary duty of the incoming winter site manager, WSM, while at the Denver office, was to assist in the selection of a South Pole winter crew and prepare them for deployment to Antarctica in late October. I arrived at the beginning of June, due to notice I had to give for the Ascension Island project. This was a little later than normal for a WSM to report, and a few crewmembers had completed interviews and selection prior to my arrival. I would work for Bill, who was the South Pole area manager. He worked full-time out of the Denver office and deployed for the South Pole summer. Bill was an icon of the USAP with many years' experience. He had entered the program in an entry-level position and worked his way up through the years. I liked and admired him immensely and much appreciated his kindness and understanding of people. He was always available to help me through the myriad of questions that I came up with. His assistant, Andrea, was also extremely kind and helpful. I came to appreciate her as the finest administrator I had ever known. She was smart, determined, and tireless. With these two, the South Pole Station was in great hands. But while they were back in the Denver office, it was the WSM that would spend the winter at the station and ensure a successful winter, or not.

My first order of business was to try to understand the massive ASC contract awarded to my new employer by the National Science Foundation, NSF. The contract included operating three major Antarctic stations, those being McMurdo, Palmer, and the South Pole; numerous field camps; and two ships. My research focused on the South Pole only, and I had little interest in the others. In reading the contract and statement of work, I noted that it was set up very similarly to what I had experienced with the military contracts I had worked on. What I found quite different was the relationship with our NSF government customer. While the relationships I had experienced up to that time with military contracts ran the gamut from extremely adversarial to civil, I had never been part of what I would call an extremely collegial partnership. This was a welcome first for me. The relationship the ASC contractor had with the NSF customer when I joined the program was extremely positive and there was great trust between us. I saw the contractor performance ratings that the NSF had given the company and marveled at the high scores the company had achieved. I had never seen such scores in any of my military contracts.

After spending some time in the Denver ASC office, I developed a great respect for the NSF and felt strongly about its mission and value to our nation. As a contract employee working for them, I was to find the collegial and mostly trusting atmosphere far more constructive and effective than the horrendous relationships I had encountered at times as a Defense Department contractor. I always felt a positive relationship made the contractor want to perform better and do more, rather than going into a defensive contractual mode, which I had witnessed so many times with past military projects.

Next, I immersed myself in the South Pole program's vast assortment of procedures, trying to understand things that were a world away and in a different type of environment than I had ever been in before. My time on rugged Shemya Island, Alaska, helped as, for the most part, vehicles are vehicles, power plants are power plants, and aircraft are aircraft, but I was now learning about a place where fuel could freeze as it reached temperatures colder than minus 86 Fahrenheit. In addition to reading the myriad of procedures during my first few months with the program, I spent much time in discussions with my boss, Bill, learning what had worked and what had not worked during various South Pole winters. There were several winters where mayhem of different sorts had occurred, and after studying that, I thought the root cause was poor selection of crewmembers exacerbated at times by poor leadership. I studied crew photographs that went back years and could put a name to the faces of rogues and some of the WSMs who should not have been there, at least not in that role. There were also years that were quite successful; major indications being the percentage of crew that chose to be photographed in annual crew photos and smiles on people's faces during photographed events, such as midwinter greetings. By taking on the South Pole assignment, I had accepted the fact that even though I had seen many things working around the world, I had not seen it all. This turned out to be quite true.

A new WSM will learn early on that the term for the South Pole is "Pole," used by most everyone in the USAP. The fact that the earth actually has two poles is long forgotten as to them, only the South Pole matters. The WSM will also learn that personnel who deploy to Pole are known in the program as "Polies." The next thing they may or may not figure out is that they have no real authority over their future crewmembers. A South Pole winter crew is comprised of various support workers, who work for several companies, and a science staff, who work for various educational institutions. The major duties of the WSMs are ensuring the health and safety of the winter crew and that the South Pole Station is operational. This means that the power plant is functioning, drinking water is being produced, food is being cooked, vehicles are being maintained, and many other things that are required for the station to be fully operational. While those duties are being accomplished, science data should be collected and transmitted. The WSM will accomplish this by persuasion and not by bully force. WSMs with military backgrounds that know only that form of leadership have a history of failing miserably at Pole. As I was somewhat of a rarity in the program, with my twenty plus years of contracting experience around the world, I had seen that type of leadership fail repeatedly at other locations. I knew that people did not respond well to a hammer, and even in the military, a true leader resorts to that as an exception not as a rule.

Every day my routine at the hotel started with a run, breakfast, and the van ride to the office. While the van ride should have been a somewhat bland experience, it turned into the sublime, with the finding of a true friend Tony, who was the hotel's morning driver. Our first couple of rides were just that, rides. Through many short rides, I was able to get to know him, and what emerged was a fascinating individual. Tony was a retired newspaperman from Chicago who had been involved in the mining industry. In his semiretirement, he was driving for the hotel and working on a very interesting and timely political book. We immediately hit it off with a shared worldview, and the short rides to the office and back every day became significant events for me. Sometimes upon arrival at the office, I had a hard time exiting the van, as the conversation was so interesting. Tony and our meeting were further examples in my life of how you can meet the most fascinating people and true friends in the most mundane circumstances, if one is open to it.

In addition to Tony, I met several other interesting people while at the Denver office. A real highlight being Mary who was the receptionist and who oversaw the front door. I enjoyed spending a few minutes with her when I happened to be on that side of the building. Many people seem to overlook or find invisible a person serving in such a capacity. I don't. Mary had been with the program a long time and knew a lot of what was going on. She was strong, opinionated, and I used to joke to the South Pole personnel, that she ran the place. I also had the honor to meet and get to know Elaine, who oversaw corporate communications. That position probably has a formal job description somewhere but Elaine with her vast knowledge and dedication to her duties was far in excess of that. Through Elaine, I was able to meet and contact some of the heroic U.S. Navy veterans that originally built and wintered at the South Pole. She was passionate and tireless in preserving their great memories and she will always have my deepest admiration and respect.

The most important thing the WSM will do while in the Denver office is to coordinate and participate in the face-to-face interviews of prospective South Pole ASC winter crewmembers. The WSM and ASC management do not interview the science personnel who will be wintering as that is done by the candidates' specific hiring institution. For the most part, that has not been a major issue as those institutions have years of South Pole winter hiring experience and have learned the traits to look for in their people. As the contract used several subcontracts to supply personnel, various company recruiters, via phone interviews, initially screen potential crewmembers. The recruiters would then rank their candidates, and those at the top were given either a primary or alternate work agreement contingent upon passing the face-to-face interview and medical and psychological exam. An interesting feature of working with the USAP is that every on-ice position has a primary and alternate person. For the South Pole winter there is one primary and there may be multiple alternates for each position. This is necessary as the primary person may not be able to deploy and there needs to be someone at the ready to take his or her place. One thing I immediately noted was that no one wanted to be an alternate.

After the recruiters had ranked the candidates, it was time to fly the primary and alternate candidates to Denver for a face-to-face panel interview. The face-to-face interview became a USAP requirement for South Pole winterovers after several less than successful winters where it became obvious that the telephone screening and hiring was not providing the program with the best possible personnel. By this time in my career, I had amassed a substantial amount of experience working on remote sites utilizing telephone interviews for hiring. I had learned the hard way of the potential problems with hiring people you had not actually seen. Out on remote sites, I had experienced several bad hires where the newly arrived employee did not seem to match the person I had interviewed via the phone. Some of the bad hires were not able to perform the job they were hired for, but most had other issues that made them unfit for the remote assignments. A few arrived not technically qualified and had exaggerated their experience, but most of the failures happened with people who were not suitably fit for a remote assignment. This could be due to issues such as excessive alcohol use or inability to work with people. It was very difficult for me to ascertain the true person by just a look at a resume and a phone call. Failures from these bad hires was disruptive and expensive. The USAP had learned the hard way the pitfalls of hiring solely by phone. They had then mandated that even with the added expense, it was necessary for candidates to undergo a more rigorous screening process prior to deployment. Now I was part of that process and looked forward to using my experience to meet and assess the candidates and be part of the final hiring decision. Out on the remote sites where I had worked, when I had a failure that could not be remedied, it was simply a termination of employment and an aircraft ride home to the persons point of hire. At the South Pole, this would be much more difficult in summer, and in winter, impossible. It was in everyone's best interest to hire the right people.

During my first year, candidates arrived in Denver and stayed at the same hotel I did. On the days of an interview, we all boarded Tony's van for the ride to the office. As I wanted to observe the candidates in the most natural setting possible, I would not identify myself as the South Pole WSM who would soon be interviewing them and just blended with the group. This gave me some anonymity and led to several amusing incidents. I would listen to banter from the group, some of which gave me an early indication as to the person's lack of fitness for the role they would soon be interviewing for. On several occasions I watched as people who had spent previous summers at Pole held court with an awestruck group who hung on every word. Some of the things being spouted were ridiculous, and I made mental notes to remember who had said what. I would identify myself to the group as the WSM just as the van pulled into the office parking lot. On several occasions, after identifying myself I heard assorted groans and comments, such as "well it's all over now." I knew of stories of previous WSMs that would anonymously sit with groups of candidates the night before interviews at the wine and beer social hour that the hotel provided. During those social events, several prospective crewmembers ended their Antarctic careers.

Upon arrival at the Denver ASC office, the prospective crewmembers received an escort that took them to a large room where they completed company paperwork and received a video presentation on the USAP and the various stations. One by one, their next stop was the interview conference room where they would meet the area manager, their hiring supervisor, the company's HR person, and the WSM. Each of the interviewers would state their position and their experience on the ice. For my first year I found that excruciating, as I had never been to the South Pole yet would soon be their leader. A question by a candidate to the panel such as "what does minus 100 Fahrenheit feel like" would get nothing but a mute stare from me, and I would awkwardly listen as another panel member would answer. I would, in lieu of ice time, state my many years of experience around the globe, but to me it rang rather hollow.

The interviews commenced with me explaining the process, and I would usually state that the questions that would be asked were not technical in nature as the candidate had already been technically vetted by the recruiter and the hiring manager. I was to learn later that in a few cases, this was not actually true, but for the most part the candidates that were interviewed were technically qualified. There were a couple of icebreaker questions, in which the candidate was asked by the panel to give a brief life history and explain why they wanted to work at the South Pole. These were important questions, and it was obvious if the candidate had done any research on what they were getting into. If they had a desire to see a polar bear, it was just about the end of the interview as polar bears only reside in the far northern hemisphere. This indicated poor or no research. On several occasions, candidates did state that wanting to see polar bears was important to them, and I can only think of one candidate who survived the interview process after making the comment. The other candidates were not disqualified solely by the polar bear comment but had other issues that probably were related to overall lack of understanding of what they were getting into. That candidate that survived the comment was given a rare pass for her faux pax as she had other redeeming qualities. She deployed to the South Pole, only lasted a week, and was sent home. It was obvious upon arrival at the South Pole that the fit was not right. It indicated that maybe the polar bear comment should have been given more weight. At least she got to personally see there were no polar bears at the South Pole.

Another interview question, which may have seemed like an icebreaker question but was quite important, asked about the candidates' off-duty interests. At the South Pole during the winter there is plenty of off-duty time. Experience has shown that the most successful crewmembers have multitudes of interests, such as exercise, learning a language, doing puzzles, reading, or other things that keep them constructively occupied through the long winter. I was slightly concerned when one of the candidates could not come up with a single off-duty interest and stated that he only liked to work and that was it. As he had previously wintered over, that answer was of no concern to the other panel members. I went along with them, and he was selected. Later I found out that, indeed, he did like to work and was excellent at what he did, and he had another highly successful winter. This was not the norm. The interview continued with a series of questions designed to ascertain the candidates' level of self-awareness and specific examples of how they had dealt with the types of situations a South Pole winterover might present. The answers ranged from the sublime to the ridiculous.

In addition to questions, there were several things the panel needed to ensure were made quite clear to all candidates. Of note is the requirement that everyone on a winter crew will be washing dishes and pots and pans in the galley's dishpit. This is on a scheduled basis, and most people understand the requirement and readily accept the responsibility. There is also a requirement that all winter personnel will serve on one of the station's Emergency Response Teams, ERT. I wanted to make it extremely clear during the interview that these were not optional additional duties but actual requirements. This needed to be done as there had been years when a handful of winter crewmembers decided that they were not being paid to do these additional duties. I always made sure candidates understood, and I would write down their responses after I had explained the duties. During the interview, most candidates wanted to deploy to the South Pole so badly that they would say about anything. I wanted to make sure they understood exactly what they were getting into and that they would be held accountable later. Another thing made clear was the need to perform all work in a safe manner. Some of the candidates had come from jobs where safety was not always held as a paramount requirement, and all needed to understand that that was not the case at the South Pole.

After many years of interviewing and the experience of observing the successes and failures of my selected candidates, I still had not developed an ironclad method of determining if I was hiring the right person, particularly for a remote assignment. For a remote assignment, you not only had to ensure you were hiring a person who was technically qualified, you had to determine how well they would handle isolation and fit into the community under austere conditions. It was all about balancing the two. During an interview, I had no great gut feel for any candidates but those on each end of the spectrum: the superstars and the truly unfit. Those were easy to spot, and I was seldom wrong in determining who they were, but the large group in the middle that could go either way was difficult to determine. This was the beauty of the panel. Having a group of experienced people on the panel who had all wintered over and knew the traits they were looking for, really helped me, especially the first year. While we followed a script, a candidate's response to any question could lead to probing questions where we might explore in more detail something in the candidate's life that could shed light on their fitness for the program.

Most candidates remember their interviews, and I found over the years at the South Pole, many wanted to discuss the interview process and seemed to enjoy hearing about amusing answers that I had received from others. There were many of those. One question revolved around a mistake that the candidate may have made while at a job and how he or she had responded to it. For most people, that was easy to come up with, but candidates were naturally reticent to bring up a mistake of any real magnitude. If they were unsure on how to answer, I would sometimes bring something up I had done, as an example. One candidate had to think all the way back to 1976, and according to him, the blunder that day was actually someone else's fault. He was not hired.

During a question about breaking a rule, a candidate brought up his stealing a car at a young age. He was hired. Several candidates exposed violent tendencies by bringing up fights they had been in. These fights escalated to the point of breaking people's bones. One memorable incident occurred when a candidate stated in response to working with a difficult employee, "I broke her thumb because she kicked me in the balls." This had occurred at his place of work, and although we continued to talk, the interview was over. Violent tendencies were something we were looking for and took any such answer quite seriously. Another answer of note had to do with how a candidate had dealt with a difficult employee. Asked that question, one fellow slapped the table and nearly shouted, "I know that one! That would be the time so-and-so tried to hit me with a creeper!" A creeper is the small platform on wheels that mechanics use to get underneath vehicles. Trying extremely hard not to laugh, I followed up with a deadpan, "Do you know why so and so tried to hit you with the creeper?" The response was almost instantaneous: "It was the drugs!" With that response and the other erratic behavior we had all noted up to that point, the interview was essentially over.

At the end of an interview, there was a segment during which the candidate could ask any type of questions they might have or provide a final statement, if they had one. Candidates could plead their case at that point, but I have no recollection of a candidate who was already a "no" in the minds of the panel members swaying us with a final statement. Their questions could be interesting and even entertaining. Younger potential crewmembers would ask about cell phone coverage and internet speed and would have a hard time concealing their disappointment when they learned that cell phones do not work, internet speed is slow, and internet is only available a few hours a day. Other common questions covered types of cold weather clothing that worked best and specifics about the station. When candidates left the room, I would usually walk them back to the conference room to give them a last chance to say anything they wanted. I would then return to the interview room where the panel decided their fate with discussion and a vote. In all the interviews I attended, there were very few times I disagreed with the panel. For the most part, with all our different experiences, we were in accord, and I think we did an excellent job of picking the right people.

The panel interview started with a candidate's arrival in Denver and did not end until the person arrived home. I always sought any kind of feedback about how candidates had behaved from our receptionist's point of view, hotel staff members who may have dealt with them, and our travel department people who interacted with them. It was important for us to know how the candidate had interacted with other people during the entire process. It was one thing to fool a panel when they wanted something, like a trip to the South Pole, but their more normal day-to-day demeanor with others was what I wanted to know about. One fellow was right on the line with his panel interview. He had a skill we very much needed but was just odd enough that we decided to reconvene and defer the final hiring decision until the next day. I was notified the next day that the fellow had contacted our travel people by phone several times late at night very dissatisfied with his seat on the aircraft on the way home. He then asked if drinks were reimbursable. If he was having a hard time with seating on a commercial flight that was only traveling a relatively short distance, how would he do with very uncomfortable Antarctic flights? The decision was then easy, and we did not hire him. What we sought during the face-to-face panel interview was to get an honest assessment of a person and try to ascertain if they could be an effective member of a South Pole winter crew. We were not just looking for superstars but more so for solid people who could perform the tasks they were hired to do, work well with other crewmembers, and not become problems.

The panel interview was only one hurdle that a candidate had to complete on their way to the South Pole. Next came a very stringent physical examination. The examination was designed to identify potential physical problems that could become major concerns while at the South Pole, especially in the middle of winter with no way out. The process was known as physically qualifying or, as we in the program called it, to "PQ." We lost many people to this process and due to medical privacy laws, usually did not know why. The PQ process started with a detailed dental examination with X-rays and for many, follow-up treatment. In the winter, there is no dentist at the South Pole. The station's physician functions in that role, but with little formal dental training or experience. Most people do not realize that a dental problem such as an abscessed tooth can become a deadly event. Years prior, while I was working out on the Alaskan Aleutians, an employee died due to complications with a dental malady.

The medical examination was very thorough and started with blood tests and a standard doctor's checkup for overall fitness. Following this was a gall bladder ultrasound and depending on the candidate's age, a chest X-ray and treadmill stress test. Candidates who did not live in the Denver area had to complete the PQ process at their own expense, hopefully with medical insurance they had, but if not, it was out of their own pocket. They would then submit the costs to the ASC and be reimbursed later. It was not an easy process for a candidate to set up all the required testing. Making it more difficult was doing this at medical offices that knew nothing about the USAP and its requirements. Another challenge was that even candidates who had alternate status and might not deploy were required to complete the PQ process. We did our best to make sure that the alternate candidate understood the value of completing the PQ process as that would open doors to becoming a primary. When the candidate thought they had completed all the required testing, they then faxed in the results to the University of Texas Medical Branch. They then waited and hoped they had not missed any of the required tests and that during the examination nothing came up that would disqualify them or require further testing. Any further testing, which was needed if something came up abnormal but was not a hard disqualifier, would be at the candidate's expense.

During my first year as part of the South Pole PQ process, a psychological examination was required. The candidate took the Minnesota Multiple Personality Index, MMPI, administered by a local Denver psychiatric office. This was usually during their time in Denver, either before or after the face-to-face interview. I had experience with this examination before my start with the USAP due to being involved years prior in the U.S. commercial nuclear power industry. In addition to the MMPI, the USAP required a face-to-face meeting with a psychiatrist who would ask a checklist of pointed questions about behaviors such as drug use, violence, excessive drinking, and other negative traits. I do not understand the value of the MMPI, and while I do see how a question such as "do you love your mother" would indicate something of a person's makeup, I am not sure how that would help select a South Pole winter crewmember. Some people do not love their mothers and have good reasons for it. I loved my mother and to answer "yes" was easy, but it might not be for others. Several questions were things like "do you like to go to parties," that seemed to be seeking out loners and introverts. The test may have also been trying to identify people whose party activities may be excessive. In general, I do not like to go to parties but answered that I did. I did not want to appear antisocial. We lost approximately 20 percent of our candidates due to them failing the MMPI. I talked to a psychologist about the value of the MMPI, and he told me that its design was to identify the truly deranged but, unfortunately, it also caught some people who were just somewhat odd. We were able to test that theory after the interview with the erratic candidate who had the misfortune of a near braining by an automotive creeper. Although he had dramatically failed the face-to-face interview and would not be hired, we still had him take the psychological examination the next day. He failed.

The entire PQ process takes time, effort, some upfront money, and a really motivated candidate to complete. At the end of it all, a candidate was either PQ'd, that is, fully physically qualified for a

South Pole winter, or NPQ'd, not physically qualified. Waivers were possible, but rarely granted. Candidates who had alternate status but had PQ'd opened the door to several opportunities, such as the ability to immediately fill in for a primary candidate who had failed some part of the process or had a change of mind. There was also the possibility of an offer for a position at one of the other Antarctic stations. Being PQ'd was a major accomplishment and was a good indication of a candidate's motivation and perseverance as it was not an easy process.

THE CREW

During my three South Pole winters, my crews consisted of forty-two to forty-six people. My first crew was the largest, at forty-six, and contained several crewmembers who were tasked with special maintenance projects. The last two winters there were forty-two members on my crew.

Scientists who work at the large telescopes take eight of the positions and have their own highly skilled machinist who can manufacture very specialized parts, if need be. The term "scientist" is used rather loosely as they vary in education from PhDs to bachelor's degrees. They also vary in experience, with some being quite involved in the specific project they will be working on, while others are hired from different industries. Their function, for the most part, is that of a caretaker, and they are responsible for doing everything in their power to keep their telescopes observing. This means they are always on call. If an alarm comes in that their project is malfunctioning, they must make a dash at any time and in any kind of weather to ascertain the problem and seek a fix. While I did not personally interview any of these people, as their specific educational institution had done that, I was happy with the selection of nearly all of them. I felt that almost all of them did a great job of performing their scientific duties while mixing with and being part of the crew. The science staff mostly made their own work schedule but were required to participate in all crew duties, such as dishpit, house mouse, and other applicable crew functions. For the most part, they did an admirable job with their project duties, and some even went above and beyond by volunteering in other areas. I well remember seeing one young man who served in a scientific role covered in dirt one day as he assisted the maintenance staff with a difficult flooring installation project. He was under the stations floor in a dirty and somewhat unpleasant, cramped environment. He did not have to be there, but he was.

The WSM is at the top of the organizational chart but in fact, does not have even one direct report as the crewmembers all work for different companies and educational institutions. With that, the WSM must lead by the power of personality and persuasion and present an image that the crew will respect and follow. During winters with WSMs who are weak or have other leadership problems, the Denver office will be more prone to running the station from afar. I very much preferred that that did not happen. It never did.

The facilities department is the backbone of winter operations. These people and their actions keep the station alive. The group is headed by a foreman, and this position can make or break the overall success of a winter, depending on their ability to lead. This is important as the maintenance people during some past winters were the cause of problems with heavy drinking, negative crew interactions, and shoddy work. I had to deal with none of those things because of the high caliber facilities people we selected and Bill J. being the leader of that group. Bill J. was an iconic Antarctic character who had worked on the construction of the South Pole elevated station years ago. He led by example, set a high bar, and was kind and caring with his men. He was tireless, incredibly physically tough, and thought nothing of going outside to find or work on something in the worst conditions. I could count on him for anything at any time. He was the closest thing I had to an executive officer, and the crew highly respected and responded well to him.

There is a facilities engineer, FE, who is responsible for the station's infrastructure. A primary duty of this position is the understanding and maintaining of proper settings on the complex computer system that monitors station services such as water and power. This system controls things such as the glycol flow, which cools the power plant's diesel engines. The heated glycol then runs through the main station and, through a system of radiators, distributes heat. The station is kept at a comfortable temperature but can change with extreme cold or warm outside temperatures and needs constant adjustment. The FE needs to have a solid background with electrical, mechanical, and computer controls. The perfect FE will work closely with the facilities staff and have the ability to respond in a civil manner to those who think the station too cold or too warm. This can be difficult to do as people run the gamut from those preferring a colder station to others who prefer it much warmer. I was lucky and had three good FEs, all different, but all 100 percent reliable and engaged. This is an important position that in certain years past had been filled by the wrong person. I was told of an FE from years past who was quite timid, seldom left his room, and smelled of mushrooms. He would not have lasted long on my crew.

There are two maintenance specialists whose primary responsibility is to walk through the station and some outside buildings and note the condition of critical items such as fuel pumps, boilers, and ventilation fans. There are literally hundreds of things to check, and the daily maintenance rounds are several miles in distance and require the specialists to be outside, regardless of the severity of the weather, which could be relatively pleasant by South Pole standards or a frigid gale. In either case, they were out there performing an especially important task. Their backgrounds can vary, but the better ones have experience with multiple mechanical and electrical systems and are experts at diagnostics and repairs. The most important traits are having keen attention to detail and being self-starters who will push themselves to actually complete the rounds rather than just filling out forms. I never had issues with that and marveled at the abilities of the men I had in these positions.

For most winters there is a plumber, although some winters the crew has gone without. The South Pole plumber's duties are much broader than what one might experience with plumbing at a more normal location. While plumbers usually deal with water and sewer lines, our South Pole plumbers could also assist with the repair and maintenance of fuel or glycol lines in the power plant. There were also other duties, as assigned, if they had completed all plumbing tasks. The station has many sinks, toilets, drinking fountains, grease traps, and other things that are in constant need of attention. Even though all crewmembers are told what not to place in toilets, obstructions are a common occurrence. I remember one that occurred at the start of a winter late on a Saturday night and our plumber trying to clear the block in the line. After multiple attempts, he found and cleared the blocked line and informed me that the blockage was composed primarily of condoms. It was quickly blamed on the summer people.

The facilities department for most winters has a single carpenter. "Carpenter" at the South Pole is a very loose term, and the common perception of a person who works solely with wood does not hold true. A carpenter at the South Pole can be involved in a variety of duties, depending on his or her abilities and motivation. These duties can range from installing flooring to assisting with the repair of an elevator. I had three different carpenters, and only one was the more traditional type who dealt mostly with wood; he was a cabinet maker. He did superb work, some of which will showcase his superior abilities for years to come. My other two carpenters, while not woodworkers, were tremendously skilled, hardworking, flexible, and worked on many projects well outside of their craft.

The winter crews would have at least one electrician, and this was a critical position. The scope of work was quite wide and could run the gamut of things as simple as changing lightbulbs to sophisticated electrical diagnostics on the station's power generators. Some crews had the misfortune of hiring electricians who were quite narrow with their electrical experience and may only have run conduit with new construction. The position needed to be filled by someone with much wider experience. I was fortunate that on my crews the electricians were closer to electrical engineers in background and flexible enough to deal with the mundane and, at times, the frightening. The frightening was the result of shoddy work that had been performed by less experienced and less motivated electricians during past winters and summers. One winter our electrician spent a sizable amount of time fixing shoddy work done years prior. Some of which had resulted in unsafe conditions with electrical wires still live that should have been deenergized and preventive maintenance that was signed off as complete but was not. Investigation revealed a past electrician that was probably doing most of his work from his desk. He had signed off on equipment checks he had not really done. A glaring example of this was a preventative maintenance job on a heat trace system that he checked off as okay and fully functional. It was discovered by our electrician during his preventative maintenance rounds that the wires were not even electrically connected.

The position of fire systems technician is difficult to fill and some years the crew goes without. We filled the position for two of my three winters and found having a fire systems technician extremely valuable. While I gave my crewmembers a choice of where in the station they wanted to live in the winter, I would unilaterally place the fire systems technician's room near an especially irritating fire alarm panel. This panel was a constant source of irritation as it would signify trouble with part of the system that was not necessarily serious and could be set off by the most minor issues. With the fire systems technician living next to it, I was assured it would get the required attention and in a timely manner. One of the challenges with this position is that the fire alarm system while needing constant monitoring does have periods where maintenance is not required. This means the crew could end up with the worst type of crewmember, one who is not terribly busy. Not being busy during the winter at the South Pole is not a good thing. Knowing this, during the selection process we looked for people who would have additional technical qualifications and would be willing to assist others. We found them.

The power plant is the heart of the station. It must live for the station to live. For it to live, it must be staffed by professionals who fully understand its operation. These people must also be able to diagnose problems and perform any necessary repairs. This may need to be done rapidly, under extremely pressured and strenuous conditions. There is a power plant lead who coordinates the scheduling of the watch-standing, which is performed by the power plant mechanic and water plant technician. The power plant is staffed twenty-four hours a day seven days a week. Nothing would get attention from the NSF, the Denver office, or the crew like a serious power plant issue. The crews understood that it was those diesel engines that were supplying the station heat that kept us from freezing to death. I was extremely fortunate to have the people I had filling the power plant positions. There were stories from years past where the program had hired less than qualified personnel and suffered the consequences. One story from a past year involved a power outage during which the station lost all power and only the emergency lights were on. During such an event, station personnel literally run to the power plant and associated electrical panels as every minute counts and the longer the station goes without power the more potential for systems freezing and subsequent serious damage. During the event, someone noted the power plant mechanic calmly sitting in the galley and making no attempt to respond. When questioned on why he was doing nothing, he replied, "It's my day off." I am happy I never had such a character on any of my crews as my reaction to him sitting there would have been quite unpleasant.

A group that was quite important, but about which I had the least understanding of their duties, was IT and communications. My initial experience with communications was as a young U.S. Marine serving in the communications section. This meant physically carrying what would now be an archaic radio and its very heavy spare batteries on a team that would call in artillery fire. That was a world away from the modern satellite systems at the South Pole. I had communications experience from all the remote projects I had worked at around the world but never fully understood the sophisticated systems and relied heavily on who I hired to run them. At the South Pole this department had four members. There was a systems administrator and a network engineer that maintained the station's internet and intranet. We also had two satellite communications engineers who were responsible for maintaining satellite communications with the outside world. Huge amounts of valuable scientific data was constantly being transmitted to institutions around the world, but what the crew cared most about was the personnel internet availability. This was at times quite a contentious issue as some crewmembers downloaded more than they were allowed. This slowed everything down for the rest of us. More than once, I was involved with curtailing excessive internet activity. In general, the department did a magnificent job of keeping us connected to the outside world.

Logistics was a group that I had issues with from my first arrival. Not the logistics personnel who worked at the South Pole, but the entire logistics system. At other remote projects where I had worked, we were heavily scrutinized by our military customers as to how we administered that program. The military knew that a poorly run logistical system would not be able to reliably supply needed equipment and materials. In such a system there would be no reliable stock of equipment or materials, and in general, the contractor could not at any given moment determine what was on hand. The government also understood that such a poorly administered logistical system could cost them extra money. In such a system, a contractor might be buying things they really did not need as those things had been previously purchased and were already there but stored improperly and lost from the system. I had seen that but never to the extent of the South Pole operation.

Upon my arrival at the South Pole, I saw what I thought was the worst logistical system I had ever seen. Adding to the problem was lack of covered storage space and many supplies being stored outside on rows of berms on the ice. Logistics seemed a seat-of-the-pants operation, where the sophisticated computer system that was supposed to track inventory from quantities to storage location and assess when to order more was not functioning as it should. The fact is, the most sophisticated computer system is only as good as the data that has been inputted, and that data was way off. What I saw was good people trying to do the best with what they had and going on "treasure hunts" to find things. A glaring example occurred not long after my first arrival, when we were not able to locate the turkeys that would be served for Thanksgiving dinner. The logistics computer system said they were there, but we could not locate them. A message was sent to McMurdo Station to send us turkeys. They looked at the computer and said we already had them. After explaining that we did not, they sent us turkeys. I believe we found the missing turkeys later, probably when we were looking for something else. This was just a minor example. It was a regular occurrence, and I cringed when I saw it. For a South Pole winter, the logistics department consists of four people: a senior materials person, two materials persons, and an inventory data specialist. During my first winter, there were a few rough spots with this group, due to the direction, or lack thereof, that they were receiving from the Denver office. But I saw improvements made every year in summers and winters and had some great people in that department. I did see an improvement to the system over my three winters.

The medical department consists of a physician and a physician's assistant or a nurse practitioner. The main intent of the rigorous PQ process is to have a healthy crew and keep the medical department from having to deal with major medical emergencies. The best candidates for the medical positions understood that if things went well, they would be virtually unemployed. During a winter free of medical emergencies, they are free to do other things, such as to assist others. Many physicians have long forgotten the art of working a mop or sponge or removing trash, but mine were all quite adept. All the medical personnel I had on my crews assisted with many other tasks, from sweeping floors to helping in the kitchen. I was incredibly impressed when witnessing them performing mundane and sometimes unpleasant tasks to help others. By virtue of their role with the crew, the physician is usually seen as a leader and an important member of the South Pole community. One of the notoriously bad winters started soon after station closing with comments about the physician being crazy. Several miscreant crewmembers exaggerated the claim and wrote blistering messages back to the Denver office on the subject. This was done in order to help gain an audience to support their mildly nefarious agenda of station control. Some of the crewmembers bought into it and a miserable winter for many of that year's crew was initiated. I am not qualified to say whether that year's physician was truly crazy but have seen photos that somewhat support the allegations.

The crew's safety engineer is an important position, and I was fortunate to have the men I had on my three winter crews. Safety is incredibly important in a place with limited medical capabilities and having a safety engineer who is active and involved with all crew activities is especially important. I had a strong background in professional safety and wanted a self-starter who I would not have to micromanage. It would be their program and I would always be there to help, but mostly to support. In hiring for this position, we looked for active people with hands-on experience and the ability to work with people. The safety field is rife with "ticket writers" who much enjoy the minor power they have over operations. There are others in the safety profession who prefer to do their work from their desk and not integrate with the workers or work being performed. I was extremely fortunate that mine were not of this mindset. For three winters we had no serious accidents and took even small first-aid situations very seriously.

The station provides meteorological data mostly to support flight operations. On my first crew, I had two meteorologists. That was changed during my next two winters to a single position with the two winter research associates, RAS, performing some of those duties. The duties of the meteorologists were to launch weather balloons, take atmospheric visual observations, and provide a local forecast. I found the forecasts to be relatively accurate, and for station operations, I was mostly concerned with temperatures, wind speeds and any major storms headed our way. The two research associates were responsible for a number of scientific support duties and were always quite intelligent and resourceful people. One thing that really caught my eye one winter was a metronome-like device I noted in their area of the science lab. It was making very concise and timed motions back and forth. I assumed it was there in support of a scientific project. Upon questioning, I found that the RAS were involved in a company sponsored health initiative that paid them a small amount of money for activities, such as miles walked. Their pedometers were attached to the metronome device, and they were thus able to log the miles from their desks! In this case, there was no real intent to cheat the health system as they were both quite athletic. It was more a case of clever fellows using their minds during a long, dark, cold winter.

The winter crew has three people who work in the vehicle maintenance facility, VMF. There is a heavy equipment operator, HEO, who does not report to that facility but does work very closely with it. The VMF has a shop foreman and a heavy equipment mechanic, HEM. It is important to fill both the positions as there is the need to always have at least two people working together for safety reasons. Filling these positions was difficult in the past, probably because the USAP ASC pay is relatively low compared to what qualified heavy equipment mechanics could make at other projects around the world. Compounding the hiring challenge in the past was the psychological examination required to medically PQ. Statistics showed that the exam had a higher failure rate for the mechanics.

Theories abound, but to me, the fact that many mechanics are quite solo in their professional practices and views might have made them more susceptible to appearing antisocial on the psychological examination. I have found many mechanics who feel that only they are competent, and I see how that type of thinking did them no favors during the psychological exams of the past.

During my three winters, I had mechanics of varying degrees of skill, from the "shade tree" self-taught to the professional factory service representative type. The VMF is tasked with fixing the heavy equipment required for winter and performing all servicing and repairs on equipment that will be used for the upcoming summer. It is important that several pieces of heavy equipment are always available as they will be used for critical operations, such as fueling the science buildings and preparing the skiway at the end of winter to land the first flight. The mechanics are not overworked and can easily finish all required tasks by the end of winter when the heavy equipment fleet should be near 100 percent operational and ready to go for the heavy workload of summer. One of the main tasks the VMF needs to be ready for is the removal of many tons of ice crystals that have blown under the station and have made minor mountains to the stations downwind rear. The incoming summer heavy equipment operators arrive on the first flight, hit the ground running, and expect the equipment to be ready.

The HEO is a critical position, and I was fortunate to have the best of the best during my three winters. They were both self-starters who needed little to no direction, and I could count on them 100 percent of the time. One of them would tell funny stories about a previous winter where the WSM had no experience with heavy equipment operations and had tried to provide direction with things like the preparation of the skiway. According to the HEO, this got to the point of the WSM bringing a measuring stick out to the skiway to check the HEO's work. It is a fact that the WSM has overall responsibility for the skiway and there is a procedure for the skiway and the measurement of any drifting, but this is something the

HEO is fully capable of assessing and correcting as needed. In this case, the additional scrutiny and mistrust of his abilities enraged the highly experienced HEO, but we had fun discussing it later. These fellows were highly skilled professionals and to do anything other than just to support them would be foolish.

The galley is truly at the center of any South Pole winter and the proper selection of the staff is of utmost importance. The staff consists of a food service supervisor, FSS, two production cooks, and a steward. To call most of these people anything other than a chef would not give them their just due. These were not fast-food cooks, and most of them had incredible backgrounds and had cooked at upscale restaurants. The food service supervisor was, by virtue of position and responsibility, an important community leader. The FSS needs to be able to keep his staff on track, and if anyone knows anything about the restaurant business, they will understand the challenges in managing restaurant personnel. At many restaurants, long hours and some stress bring drugs, excessive alcohol use, and an unhealthy lifestyle from socialization and lack of sleep. While the galley hiring company and the panel interview try to ensure the best hires, every now and then a person with problems slides through the hiring process. Sometimes the problem is the FSS, but historically, it is more likely to be the cooks, which the FSS then must get through the winter. I was fortunate and only experienced minor issues with galley staff.

Another especially important facet of the FSS's responsibility is the fielding of complaints from the station's food critics. Personally, I had little time for complainers, except on a few occasions, and I always wanted the FSSs to know I supported them. There were at times valid complaints that were more suggestions in nature, and a good FSS will work with the community members for those. At other times there were complaints that had little to no validity, and I remember one that stood out that stated "steak was a little cold." I had never personally experienced that, but what bothered me most was that I knew the complainer lived in a shack in Alaska and probably ate beans out of a can. At one point, there was a ranch-style, triangular dinner bell that was labeled the "whiner bell." I saw it rung a few times for a complainer, after my first arrival. At some point it fell out of favor with the company that ran the galley and was removed, but I personally found it appropriate.

The steward is the real unsung hero of a South Pole winter crew. The steward works in the dishpit for breakfast and lunch five days a week and does station janitorial duties on Saturdays. The steward is also responsible for the overall operation of the small store, although volunteers will assist. Stewards always start the winter extremely motivated and excited, and then, with the repetition, drudgery, and hard work, they do get worn down. The crew usually sees this and will help by taking turns in the dishpit and cleaning the galley. During my first winter, we unexpectedly lost our steward at the end of summer, just before the station was to close. A quick replacement was found in McMurdo, who had little time to consider what he was doing but made a quick decision to join us. When that aircraft arrived, I saw him as a VIP. I do not think the crew really understood the significance, but if we had no steward for the winter, the crew was going to be spending a lot of time in the dishpit. At the heart of the most successful South Pole winters will be a galley staff that can cook great food, get along well together, and interact in a positive manner with the community. This is not always easy to do.

There was another type of crewmember. One that would never journey to or experience a South Pole winter but was an essential element to its success. That group consisted of the mothers, fathers, brothers, sisters, spouses, children, and other loved ones in a winterover crewmember's life. While not physically at the South Pole, one could feel their presence through their associated crewmember. Some of these people had sacrificed, and many of them worried about their loved one spending a winter at the South Pole. My wife, Melissa, who was a veteran of several overseas remote assignments, understood quite well what I was doing. I had been gone before but never for quite as long and always with the ability to return, if need be. Through the years while I was away, she had faced many things, mostly mundane but a few truly serious. During my three winters at the South Pole, alone she would face hurricanes and a pandemic.

The selection of the right crewmembers is of extreme importance in how a South Pole winter will go. The perfect crewmembers would be extremely proficient in the field they were hired for, and they would love doing their jobs. They would also be able to peacefully coexist with all types of other people with extremely diverse backgrounds. They would have no extreme negative traits, like problems with alcohol or drugs. In addition to those things, there was another trait that I always sought and sometimes got from my remote work around the world. They would have other skills. It is not impossible to get a mechanic who also can operate a crane or a carpenter that can weld. When putting the crews together, I was not just looking at what would be the challenges of a normal South Pole winter. I was looking for people that would effectively respond in the event of a major disaster, such as the ice shifting and the station breaking in half. While the chances of such an occurrence were extremely remote, I wanted people who were ready for such things. The South Pole is a very extreme place, and I wanted extreme people.

Once a candidate had successfully passed the face-to-face interview, was fully PQ'd for a South Pole winter, and (during my first year) had passed the psychological examination, they were now crewmembers. The WSM working from the Denver office was responsible for contacting all crewmembers and ensuring they were on track for deployment. During these contacts, questions came up about all types of things, such as what type of facilities existed at the South Pole, what things were available for purchase at the small store, and, most commonly, extreme cold weather clothing selection. The most important thing that occurred during this time was the crewmember's placement in either fire or wilderness first aid training courses. I would seek any kind of past fire or medical experience that the crewmember's may have had, and if they had none, ask what interested them. The crewmember was also made aware of the need to participate in a teambuilding session that occurred either before or after the emergency training.

All crewmembers are required to participate in one or the other emergency training classes prior to deployment and are expected to serve on the ERT after arrival at the South Pole. Most people want to attend the one-week fire school conducted at a local Denver fire academy. Crewmembers on the fire team are responsible for responding to any type of station emergency, and in the event of a fire, they would be the team to extinguish it. At the South Pole there is no calling 911, only a response from a small group of dedicated volunteers who have gone through a one-week fire brigade training course. During the interview process, I always asked if the candidate had prior fire experience as personnel who were U.S. Navy veterans or National Oceanic Atmospheric Administration, NOAA, crewmembers who had been on ships usually did. There were also people who had been volunteer firefighters, and this background was quite useful. Early in the process, I was looking for potential team leaders, and anyone with some experience was a strong candidate. A smaller group attended a wilderness first aid course conducted by a private provider. These people would assist the South Pole winter physician, when dealing with a medical emergency.

Fire school started with the group of crewmembers coming together at a nearby hotel and, if they had not yet been to teambuilding, meeting for the first time. It was my first chance to see them in person and note their behavior under stress. Most crewmembers did not know that they were still involved in the crew selection process, which would go through their deployment training and summer at the South Pole. Some of them might not be deploying as contracts could be ended due to negative traits being noted during the training. Others might deploy and be sent home during the South Pole summer season, prior to the start of winter.

During the face-to-face interview process, I always kept in mind that most candidates wanted to appear in the best possible light. We had something they wanted, a trip to the South Pole. Many of them would say nearly anything to try to impress the panel. I always kept in mind that during an interview in a nice Denver meeting room, I was seeing the candidates at their best. Those same people might be quite different after months of isolation, darkness, and cold, long after they had taken their hero photographs, proudly standing at the Geographical South Pole. During the fire and medical training, teambuilding, and the deployment itself, at least some of the real person would become visible. It was not always what we hoped to see. On the eve of fire school, I greeted one of our new crewmembers shortly after his arrival at our Denver hotel. He had only been at the hotel a couple of minutes and was already at the bar with several beers and shots in front of him. He was a genuinely nice fellow but did not deploy.

A Denver fire academy conducted the one-week fire brigade training. The USAP had used several different training institutions through the years. During my time, we used the Aurora, Colorado, training facility, and it was first class, with top-notch instructors and excellent fire-training capabilities. The school started with a classroom overview of fire, personal protective equipment, and various methods for extinguishment. Following that, each student was issued fire department bunker gear and given an introduction to the self-contained breathing apparatus, SCBA. I had experience with the SCBA from my years of hazardous materials response out on the island of Diego Garcia in the Indian Ocean. At the fire school, I must admit, I felt its weight more with my age. The academy staff were superb in their approach with our groups and understood well that we were a volunteer fire brigade and not a professional fire department. They gathered information as to the South Pole Station's layout, firefighting resources, and gear we used, and they tailored their instruction to those things. They pushed us in the training but did not try to push us to any kind of breaking point, as they would have done to a candidate in one of their professional firefighter classes. The academy's professionalism and our class's motivation and effort were quite impressive. The wilderness first aid course was shorter, but it gave the students a solid background in advanced first aid so they would be more capable of assisting the station's physician during an emergency. I was not as concerned by its outcome as our physician could train the same thing while at the South Pole.

For my first year, teambuilding came first, and fire school came afterward. For my second, the order reversed. After experiencing both, I found that attending teambuilding first was the preferred order as it allowed me time to learn all names and something about each crewmember prior to the physical stresses of fire school. It also gave me initial information as to who my potential fire leads could be as I could look for any leadership traits during teambuilding. I could then observe them in a leadership role during the fire school's hands-on activities. After that was completed, I would have a good idea as to who my fire team leads would be prior to arrival at the South Pole. While preparing for my first winter, the morning that fire school was to start, I conducted a roll call in the lobby at the hotel where we were all staying. We had a missing crewmember. I called his hotel room and received no answer. With concern, I went to his room and had the hotel staff open his door. He was gone! There was a note left as to why he felt he could not be on our crew. He wrote that he liked to be alone and after the many group exercises and interactions he had experienced during our teambuilding session, he felt he would not fit in. I wish I could have talked to the fellow prior to his fleeing as I could have told him he would have plenty of alone time at the South Pole.

After completion of the fire school classroom training and the issue of bunker gear, the crew experienced fire—real fire! One of the first exercises was breaking into small groups and entering a concrete building that was set up as a fire chamber. Inside, a very controlled burn was taking place. The heat rose to the point that it was much hotter than any oven, as we sat silently in semidarkness watching the flickering flames and smoke emitting from a pile of wooden pallets burning in the center of the room. It did not take long to start feeling the heat through the bunker gear. We learned quickly that even flexing the sleeves on the coat could cause a hot spot on your arm. We also noted that it was much hotter the higher you were and staying low was a good thing for both heat and visibility. While crouching on the floor in the concrete fire chamber as it got hotter and hotter, I really had to concentrate and fight the urge to panic or, at least, the strong desire to exit that room. After we cooked to the satisfaction of our fire training instructor, we departed the room and building to our great relief. Unfortunately, a few minutes later, we were back in the chamber as the training staff had been able to raise the temperature a bit more and wanted us exposed to that.

After our thorough broiling in the fire chamber, while checking time, I noted that the crystal in my Rolex diver's watch was now opaque. With further investigation, I saw a small section of the rubber O-ring, which should not have been visible, exposed on the back of the watch. Somehow, the seal had been broken and steam from my perspiration had entered the watch. It was a costly repair, which I was not able to do for over a year due to our immediate deployment south. The next year, I made an announcement to my crewmembers to remove all jewelry prior to entering the fire chamber, and I did the same.

The crew practiced making entrance to a building with smoke and flames while searching and then rescuing any occupants. This could be training dummies or other crewmembers staged as victims. It was very realistic training, and after a couple of days of that, we were becoming proficient. After our few days of training, we all developed an immense respect for the fire academy staff and firefighters in general. With the conclusion of fire school, our group would become the only firefighting force at the South Pole Station during the long winter. It was a sobering thought.

Teambuilding was three days at the Estes Park YMCA doing various exercises with an experienced trainer knowledgeable in the development of positive group dynamics. Shawn was that and much more. While initially skeptical of the teambuilding process due to preconceived notions of silly group hugs and trust falls, after meeting Shawn and experiencing the first day of his program, I became an advocate. Shawn's teambuilding experience did not come from kids' summer camps. He had developed and completed serious projects seeking to find common ground and cooperation between incredibly disparate groups. A major project he had completed that few would consider tackling was between Israelis and Palestinians. Shawn was no lightweight! Teambuilding starts with a beautiful group drive from Denver to the Estes Park YMCA facility, nestled among snow-covered peaks and covered in pine forest. It really is a fantastic drive with beautiful scenery and elk herds that make that area their home. During the drive, I was able to gather interesting information from crewmembers who wanted to drive up with me. One was so talkative that I made sure she rode the return trip in another vehicle to give her new crewmates the chance to experience her many thoughts on everything and anything.

Teambuilding commences with exercises designed to get the group to get to know each other and then effectively interact. One quote that I genuinely believe in came from Will Rogers, who said, "I never met a man I didn't like." I believe that is true with all but the most damaged individuals. The key is to spend the time and effort to really get to know someone. Once you understand the true person, you can make allowances for their behavior and understand why they act the way they do. Actions or words that one might perceive as offensive may not really be, when consideration given as to what made the person the way they are. Getting to know someone better is usually a good thing, and the teambuilding was a real crash course in that. Upon arrival, the crew checked into rooms in a large lodge and had time to get out and explore the beautiful mountain area. Formal classes then started and consisted of many exercises, both indoors and outdoors. These exercises encourage and teach people who have never met to appreciate, trust, and work together to achieve goals. The crew is being closely scrutinized during the program, and on several occasions, I saw several things that caused me concern. One was an interesting event that occurred when I had a crewmember inform me that two of my crewmembers were not getting along. After looking into it, I found that one crewmember strongly disliked another. The disliked crewmember probably did not even know that. The crewmember then told me that a small group of them had tried to intervene and at dinner one night made suggestions to her about trying to talk to the subject of her dislike. Her answer was "I am so done with so-and-so!" This is the type of thing that teambuilding tries to prevent, as in a long South Pole winter, it is much better to be able to work toward ironing out differences than just being "done" with someone. On the other hand, the fact that the crewmembers tried to intervene and then, when unsuccessful, came to me, are things that teambuilding promotes. One of the great things that teambuilding promotes with several exercises is for people to try to work out differences among themselves, which will entail understanding other viewpoints. It takes people to be willing to do so, and some people have a hard time with that. At teambuilding and fire school, many lifelong bonds of friendship are sealed as people start to bond and really get to know each other on the eve of a major event in their lives—deployment to the South Pole.

At the conclusion of required training, the winterover crew divides into those deploying immediately to the South Pole and the experience of a summer and a winter. They will begin arriving around the first of November. The other group deploys later, solely for winter, and will arrive in late January or early February. Goodbyes are said between these two groups, with the expectation of meeting up again at the start of winter. For the most part, this turns out to be true, but there are also rare cases where winterovers do not survive the summer and are sent home. I experienced that my first summer, and it did not happen the next.

It is a long flight from the West Coast of the United States to Auckland, New Zealand. I remember on my first flight hearing a conversation between one of my crewmembers who had never traveled internationally and his new seatmates. The discussion was mostly concerned with him and the amazing experience he was to have at the South Pole. This went on for some time, and it sounded like lifelong friends had been made, but as most experienced travelers know, that would probably be the last contact between them. During my first deployment, I sat in silence and pondered what was to come. I was in new territory now, heading into a world I did not know with a group of people I barely knew, to spend a year as their leader.

After arrival in Auckland there is a transfer to the domestic terminal and a flight to Christchurch, where the USAP operates the Christchurch Deployment Center, CDC. Upon arrival in Christchurch, a USAP representative meets the group and provides transportation to the various hotels the program uses. It is a highly organized process and far better than most of the military projects I had worked on, where we were more on our own. The USAP realizes that some of the crewmembers are not necessarily experienced world travelers (although many are) and try to make the experience as easy as possible. After finally reaching their hotel rooms, crewmembers are usually exhausted from the long trip and have at least a couple of days to rest and get into the new time zone. Sometimes, due to weather or aircraft problems, crewmembers spend days and even weeks in Christchurch. Veteran crewmembers rarely complain of this, but those new to the program are anxious to get to Antarctica and the start of their adventure.

Christchurch is a beautiful city located on New Zealand's South Island. It has a population of around 380,000 people but to me feels smaller than that. The city suffered a devastating earthquake in 2011 and a decade later, the effects can still be seen. The city has a wonderful botanical garden that contains the Canterbury Museum, with its epic historical Antarctic displays. Out near the airport is the

International Antarctic Center, which has incredible displays and at times even exhibits beautiful live sled dogs. The nearby harbor at Lyttleton was the last major jumping-off point for several early Antarctic expeditions. I much enjoyed my time in Christchurch, especially after completing a South Pole winter. During the stays after being at the South Pole for a year, I would enjoy the class and hospitality of the Heritage Hotel, which was built in 1913. The hotel exudes charm and elegance from another time. Being in Christchurch after experiencing relative deprivation for a year and being able to walk in warmth and buy and eat whatever I wanted was a sublime feeling. Going through Christchurch on the way to Antarctica at the start of a year was a good time, but I was always keen to continue the journey south and did not enjoy delays.

The day after arrival, the crewmembers receive transportation from their hotel to the USAP CDC where they receive a briefing on the program, get flu shots updated, and are issued extreme cold weather, ECW, clothing. The USAP issues everything that a crewmember will need to protect themselves from the harsh Antarctic environment except for thermal underwear and socks. The classic "Big Red" parka issued with the USAP patch and nametag attached is warm enough for the coldest days at the South Pole, where temperatures can exceed minus 100°F. I only wore the Big Red for the flights, as it was a requirement, and at the South Pole, I had my own military ECW, which was lighter and more flexible. Later, I would acquire two anoraks with histories.

The ECW issue is a bit mystifying for the new USAP members as there are some clothing choices to make, and what one needs is very dependent on their station, either McMurdo or the South Pole. In addition, it matters if the person is going down for a summer or a winter. Summers are relatively mild, but winters are not. Fortunately, there are usually veterans who will help with advice and there can be minor changes once the person has arrived on station. For my first year, my initial issued boots were not warm enough for the South Pole winter walking I was doing. Fortunately, I found a better boot option in the station's supply system.

Once a crewmember has received their ECW, it is stored in orange bags at the CDC in the changing room. If things are on schedule, the next day they arrive at the CDC early, weigh their bags, attend a final safety/environmental briefing, and board the aircraft to Antarctica. The new people are easy to spot, with their enthusiasm and unfamiliarity with the process. The old hands are usually more sedate as they have "been there done that," for some, many times. Flying times vary from Christchurch to McMurdo Station, depending on if one is on a C-17 jet or a C-130 with propellers. I have been on both and much prefer the speed, legroom, and toilet that a C-17 offers. For many new to the program, this was their first time on a military aircraft and quite the novel experience. For me, it was something I had done many times as a contractor and had even parachuted out of C-130s while in the military. The professionalism and attention to detail of military flight crews always greatly impressed me. I knew we were entrusting our lives to them and those beautiful U.S. aircraft, as we flew over treacherous Southern Ocean waters. Prior to takeoff the military crew chief conducts an abbreviated emergency briefing as one would get on a commercial aircraft. While it sounds reassuring to know where the life jackets and rafts are located, the harsh fact is that a crash in those savage waters would not be survivable.

Upon arrival at McMurdo, the passengers are loaded into several types of vehicles, depending on their number. If there are only a few, a van will be used. For a fully loaded aircraft, the iconic large red terra bus, which seats many people and is very slow, will be used. The drive to McMurdo Station is scenic, with views of the nearby mountains, frozen sea, and ice, a lot of ice! Most of the newly arrived are somewhat overwhelmed by the experience, and for some, it is the fulfillment of a lifetime dream. Cameras snap constant photos of anything and everything. I always sat in silence and experienced no joy. I was on my way to the South Pole, and this was simply a minor diversion.

There are many people who love McMurdo, but initially I was not one of them. For my first two trips through, I felt the best thing that could happen to me at McMurdo was a quick transit in either direction. McMurdo gets described most as resembling some type of Antarctic mining town in physical appearance. I fully agree with that, but once one enters the 155 Building, where the galley and other central McMurdo operations reside, another aspect of its character emerges. This character is based on at least some people in the workforce that exist solely for the next social activity. At these social activities, they can display the expected Antarctic antics perpetuated as people try to outdo each other with fake irreverence. I personally love the idea of real irreverence, which was something that started institutions such as Mardi Gras. At McMurdo, an attention seeking, drunken, male vehicle mechanic wearing a tutu at a social event is nothing unusual. I know this does not apply to all McMurdo residents, and I have met many who did not fit into that mold, did a great job, and appreciated McMurdo for its other features, not the silliness. I do not know how they can coexist with some of the ridiculous antics. For me, early on, the exit from McMurdo in either direction was always the best part. Later I developed a better understanding and appreciation for the place and all the great things that are done there, which softened my view.

The flight from McMurdo to the South Pole is an amazing event. Being the WSM and needing to get to the station as soon as possible for season opening, I always flew on a DC-3, known as a Basler. These aircraft make several trips to the station prior to the military LC-130s. The LC-130 is a military C-130 but has skis for landing on icy Antarctic skiways. The Basler's are beautiful and truly vintage aircraft with their airframes dating from the 1940s. These aircraft have been updated with the latest electronics but are still simple, tried-and-true. They can land on about anything and are polar workhorses. The Basler's route is limited in altitude and thus there can be a fantastic view of the Beardmore Glacier that Earnest Shack-leton discovered, followed by Captain Scott a couple of year later. At the top of the glacier, the aircraft continues the last few hundred miles on the polar plateau and the new arrivals may start to feel the lack of oxygen as they have gone from sea level to over nine thousand feet. Oxygen masks get used by some, but for whatever reason, I never needed one. After several hundred miles of nearly level flying, the aircraft arrives at the Amundsen-Scott South Pole Station. This station—a ship of sorts that rides atop the moving ice—was to be my home for almost three years.

THE SUMMERS

Summertime at the South Pole runs from the beginning of November to about February 15. During this time, the station's population swells to its maximum capacity of around 150 people. The current South Pole elevated station was completed in 2010 and replaced the iconic dome structure that had served for many years. It is a beautiful design and, I think, a work of art. The station sits on giant pillars sunk down in the ice that can be raised as the ice level around the station rises each year with new accumulation and causes the station to sink. The station is elevated, faces the prevailing winds, and its aerodynamic shape allows the wind and ice crystals to pass under the building. This prevents excessive drifting to its windward side. Drifting does occur to the station's downwind side. During the short summer, the drifts are removed with the aid of heavy equipment. Directly under the station is a hard icy surface, in some places resembling an ice skating rink as the relentless winds have scoured it. Having the capability of raising the station on its massive support pillars will allow it to stay above the new ice whose level rises at about a foot a year. Being able to be raised, it will serve as the U.S. monument to its interest at the South Pole for longer than the initial station built in 1956/57 and the later geodesic dome built in 1974. Both were eventually overcome by rising ice levels, buried, and had to be demolished. The current station, known as the elevated station, is approximately fifty thousand square feet in size and includes a galley, greenhouse, gym with basketball court and weight room, sauna, arts and crafts room, music room, library, several lounge areas, and 150 one-person rooms. It is a vastly different experience from Roald Amundsen and Captain Scott's tents.

After my first arrival and during a discussion about the upcoming winter led by a visiting NSF representative, I was introduced to the assembled group as "the captain of the ship." During my years at the South Pole, I found that title accurate as there are nautical aspects to the station and the position itself. The station sits on nearly two miles of ice that move toward the Weddell Sea at around thirty-three feet per year, the station is like a ship as it is self-contained, with power, water, and food, and it moves with the frozen sea of ice. There is a "crew" who eats in a "galley." There is an "observation deck," and some of the interior windows are of porthole configuration. In the direst of emergencies, a wing of the station can be sealed off and will function as a "lifeboat." The station and crew, especially in winter, require a captain of sorts, and an effective WSM will function in that role.

On my first arrival, I climbed down the short Basler ladder and immediately felt the temperature of minus 58 Fahrenheit. I had never been in such cold as that. Now, with ice and not land beneath me, I took my first steps and looked around in awe at the extreme whiteness of the frozen sea I stood on. It seemed to go on forever, devoid of any large natural features. It was ice, hundreds of miles of dead ice in every direction. Ice I did not yet know or understand. While lingering around the aircraft helping the other crewmembers assembling baggage for transport, someone told me the tip of my nose was turning white and I should get inside the station. It stung a little and surprised me with how quickly it happened. This was my first minor lesson in how extreme cold affects the human body. It would not be my last cold weather learning experience, and over time, I would become quite proficient at dealing with it. That first day it really caught my attention and scared me.

I arrived at the South Pole with a vision and a plan. First, I would need to know and understand the station itself. I needed to know how things physically worked and where any vulnerabilities might be. Initially, I was most concerned with the state of the power plant as that had been so important out on other remote sites I had worked. At the South Pole it became much more important as it would keep us from freezing to death. There had been power reliability issues during the winter prior, while my crew was being formed. I would listen to phone calls between the power plant people at the South Pole and the Denver technical experts and at times found it disturbing. The South Pole power personnel were struggling with generator units that would suddenly drop off-line, and they did not always know why. Making it worse was the fact that the technical wiring diagrams were not accurate and up-to-date. During the phone calls, you could sense real stress from the South Pole personnel. They were being advised to do things to fix the problem, but they were seeing wiring that was different than what the Denver personnel were seeing on their diagrams. They had a bad computer system that regulated the units and ended up spending much of their winter running the generators in a manual mode. They struggled through the winter, and I really wanted our winter to be smoother. Fortunately, that summer, major repairs were initiated, and while we had some power issues during the winter, it was not to the extent of the previous year. The power plant was incredibly important, but it was only one system of many. I needed to have a solid understanding of them all.

Secondly, I needed to understand the crew and how I could best interact and lead them. I knew I would maintain a distance but would have to determine how wide the gulf would be. I had learned that the distance should be enough to maintain a level of respect but not so wide as to appear uncaring and aloof. By this time in my career working at remote sites around the world, I knew I had not seen it all. What worked out on a tropical island or in a war zone might not work here. These people were different than most people I had worked with on military projects around the world. With the Defense Department projects, there were usually many military veterans on staff. The projects had a sort of quasi-military organization and feel. While I had witnessed that they did not respond well to true military leadership, they did understand a chain of command of sorts and responded to it. During a crisis, they could revert to a military mindset and tackle challenges that way. During the interview process that put the South Pole crew together, I had few military veterans and quite different types of people. They were more eclectic and free thinking. While unsure of some of the specifics, my early plan was to lead by example and promote the concept of a crew. A crew that for each year would become an entity that could never be repeated. I wanted to foster the concept that what people did during their winter mattered and counted. It was permanent. I wanted the crewmembers to be proud of their individual and crew accomplishments. I wanted all of this and to accomplish it in an atmosphere that understood and respected those that had gone before us.

After arrival at the South Pole, I immediately met the WSM that I would be replacing and was quite impressed by his level of professionalism and dedication to duty. He was retired military, and it showed. Unfortunately, the military mindset does not work well in the USAP unless under the direst of emergencies. He had faced a lonely year with a crew that marched to a different drummer and some who openly antagonized him with antics that would seem funny if they were done anywhere but over a South Pole winter. The winterovers we would be replacing were pale and shaggy, and I envied them tremendously. They had all done what I had not, they had successfully wintered at the South Pole. They were now members of an exclusive club that at that time numbered only 1,500 people. Over 5,000 had summited Mount Everest, most of them paying customers. There was a different price to be paid for a South Pole winter. I attended the ceremony where they received the Antarctic Service Medal with winterover clasp and was truly humbled as they each received the medal and a few words said.

Standing in that room and feeling the strong bond that had been created between them was a remarkable experience. I was happy for them and wanted so badly to experience it myself. I was to be at similar ceremonies for three South Pole winter crews that I was to lead, and each, for me, was always slightly overwhelming.

My initial room assignment was in the "B" wing. In that wing, the rooms are slightly smaller and close to the B-1 lounge. For some people that is an advantage and to others a nuisance. After a couple days in that room another became available that I wanted. That room was at the end of the hallway near the outside exit doors, B-219. I was to spend three winters in that room. The B rooms are small, approximately eight feet wide and twelve feet deep. The rooms have either "hard" or "soft" walls. The soft walls are modular and can be removed to form larger rooms if there are couples. My room had a soft wall, which was not fully sealed. I thus became intimately knowledgeable of my neighbor's habits each summer. The rooms that are located along station exterior walls have windows and are known as outside rooms. The outside rooms are usually preferred, especially by new summer people who do not fully understand the impact of the twenty-four hours of daylight and the fact that sleep may be difficult. There are also inside rooms that have no windows, and a few people, usually with sleeping problems, will ask for them. Most people want a window, but for winterovers, the room's windows are covered after the sun sets so it does not really matter. All rooms in the B wing have a small desk that faces the window, or the end of the room if it is an inside room. There is a small twin bed and a wardrobe cabinet. My room had a bed raised approximately five feet off the floor. To get in it, I climbed a beautiful, wooden, curved, shop-built ladder. This ladder had been built at the South Pole and I always marveled at the craftsmanship but never knew who had built it. On the bed, I placed a light with a flexible cable set up for reading. This arrangement gave me room under the bed to store things, such as all the boots and shoes I had as well as my large canvas travel bags. My small room was remarkably like a cabin aboard a ship, and I loved its coziness and effi-ciency. My home in Texas was filled with exploration and exotic items from around the world. It had become so packed with such things that I had bought the home behind me as a guesthouse. At the South Pole, my world was quite compact.

The summers are a busy time at the South Pole. The weather is relatively warm and will at times get above 0°F. The sun is up for twenty-four hours. During this time, any major construction that needs to be performed occurs and scientists swarm the station along with gaggles of graduate students who serve as labor. Science at the South Pole consists of four major projects. The Ice Cube Labora-tory, ICL, is a true ice cube that functions as a neutrino observa-tory. It consists of a cubic kilometer block of ice below the surface, which has around 5,200 digital optical modules, DOMs, embedded throughout. This massive subsurface ice cube can detect incoming neutrinos. The neutrinos are products of cataclysmic events that have occurred throughout the universe and can be traced back to their point of origin. The South Pole Telescope, SPT, is a tele-scope with a ten-meter dish that is utilized to detect emissions from cosmic microwave background, CMB. This is used to find distant galaxy clusters. The SPT is a collaborative organization that con-sists of several other scientific organizations located throughout the world. The Martin Pomerantz Observatory, MAPO, is another astronomical observation system that has been the site of several very prominent projects. As technology has progressed, the sys-tems have been upgraded and utilized to support astronomical projects, from detecting neutrinos to measuring CMB. The Atmo-spheric Research Observatory, ARO, is run by the National Oce-anic Atmospheric Administration, NOAA. This project measures greenhouse gasses, stratospheric ozone, and many other things. It is part of the global atmospheric monitoring community, and the data is used to try and understand things like climate change and its root cause. In addition to these major, year-round science proj-ects, during the summer there is a frenzy of other scientific activity, with other types of research being performed. One project I found fascinating was the SPICEcore project that drilled down more than a mile into the ice. Ice core samples that dated back thousands of years were extracted and scientists were then able to assess ancient atmospheres.

According to some winterover personnel, there are four types of people and a hierarchy of sorts at the South Pole Station during the summers. The first, and at the top, are the winterover personnel who are also there for the summer and will be part of the station's emergency response team, ERT. They will be staying a year at the South Pole and look at things with a long-term perspective. They see the station as a home rather than a motel room. Second are the "summer people" who deploy from the Denver ASC HQ or scientists who come from various institutions. They remain on station throughout the summer and from the first day of arrival, know approximately their departure date. They are usually making their travel plans home far in advance of that. Some of these people have been making the trip down for years, and a few have wintered in previous years. Some of these people see this as an escape from their daily drudgery and use this time for antics they might not do at home. Next come the "visitors." These people are on specific short-term assignments that might range from a single day to a couple of weeks. The bottom of the hierarchy is the "tourist." They pay fortunes and arrive by airplane, vehicles, skis, and even bicycles. They are mostly ignored and somewhat disdained by most of the station staff, and other than some getting station tours, tourists are confined to a nearby tourist camp and, thankfully, quickly depart.

During my first summer, I concentrated on learning about the station's infrastructure. Of primary importance was the main power plant. The power plant has three Caterpillar diesel generators rated at about 750 kilowatts each. There is also a small "peaker" unit rated at around 450 kilowatts. These units supply all the needed power to the main station, the science projects, and other buildings. They are the heart of the station and were to me, with my years at remote sites, of extreme importance. Understanding that importance, the first note I made in my notebook after arrival at the Denver office concerned power issues then occurring at the South Pole. The stations water comes from a well system known as a "Rodwell." The name comes from the fact that a very smart fellow named Rodriguez developed the simple yet ingenious concept. Warm water is injected down deep into the ice and a bulb of water forms. This water is pumped to the station, where it is placed in water tanks. Although quite clean, it is then sterilized for drinking purposes. I enjoyed South Pole water and soon learned the location of what I thought was the best drinking water fountain in the station, where I would obtain my water. As the climate is so dry, the drinking of water is highly encouraged, and most people are not far from their water bottle. The B pod is the station's "lifeboat" as it has large freezer doors that can be shut, which effectively seals it from the rest of the station. This lifeboat can contain a winterover crew and would only be used in the direst of emergencies. It has a fully functioning emergency power plant, the ability to make water, a small kitchen, and even a washer and dryer. To date, it has never been used, and I hope it never sees service. If it ever does, it will mean the rest of the station is effectively cold and dead.

The summer construction projects take advantage of the twenty-four-hour-a-day light and the relatively warmer temperatures. While South Pole summers are cold by most standards, they are not that much colder than a North Dakota winter. This brings an influx of construction and maintenance people who are quite different from the scientists. It is an interesting mix, and really, they do not mix all that well. Many of the science people are socially awkward and tend to stay with only their own small groups. The construction and maintenance people are more gregarious and tend to be able to cross boundaries, with most having an interest in the science that is occurring and a major interest in the South Pole summer social scene. It was an interesting mix of people, and I saw all kinds. I remember one summer person who stood out; he was a complainer. He complained about one of my winter crewmember's T-shirts that had the outline of a rifle printed on it. This delighted my winter crewmember, who did not hold back as to what he thought of the fellow. I do believe he may have worn his rifle t-shirt more often that summer. The complainer then focused his complaints on the galley and the music they played during the time between meals. This delighted the galley supervisor, who then played the song "Afternoon Delight" over and over while the complainer ate. At one point, many in the galley were singing the silly song because after hearing it so many times, they knew all the words to it. The complainer had issues with a few other things that summer and was the kind of person that should never be on a winter crew.

During the summers, there are many social events and most involve alcohol. I have found alcohol to be the bane of most remote assignments, and the South Pole is no exception. During the selection process for winterovers, candidates face scrutiny for signs of problems with alcohol. This starts with a question during the medical evaluation asking how much alcohol the person drinks. The challenge is that most people answering that question will usually say they drink less than they do. Lessons learned through the years indicate that when a candidate says he or she has X number of drinks per week, that actuality translates to that many per day. The winterover crewmembers are scrutinized for signs of problems with alcohol during the period after the formal selection process. This will include pre-deployment training, deployment itself, and for most, a summer at the South Pole prior to winter. The summer people are less scrutinized due to their short stay and the fact that they can be sent home at any time.

The summer was a good time to evaluate personnel for winter fitness, and there were epic meltdowns by some. One winter crew that was not mine lost over 10 percent of their members due to summer antics. One of their most prominent meltdowns concerned a fellow who stated to others that he was a "human weapon." He was able to, in what he later claimed was blackout drunkenness, threaten and punch a couple of station residents and slap a female on the ass. His winter contract was quickly terminated, and he found himself placed on the next aircraft out. In his case, and much to the station residents' dismay, someone saw in the fellow's hometown newspaper a story based on an interview with him. In that story, he detailed his heroic efforts while at the South Pole and left off anything to do with his early departure. I believe station residents sent in a couple of rebuttals to the newspaper. That crew also had the honor of having another winter crewmember losing his job during summer for being so drunk as to pass out on the floor of a restroom, pants down and shitting himself in the process. He was discovered in that state and sent home.

After the first summer, when I lost one good fellow due to his drinking, I became much more involved in evaluating crewmembers and their habits with alcohol prior to deployment. I also began to accept information that came from technologically savvy and concerned crewmembers who would perform internet searches on new crewmembers, specifically self-posts on social media. One newly selected crewmember had repeated self-postings on a social media website in which he stated many times that he was drunk and bored. Sometimes, this would include photos of him that certainly supported that statement. For a South Pole winter, this was not a good combination. Interestingly, when I forwarded this information to his employer, there were some in his management who were more concerned that we were cyber stalking than with their employee's public online persona. As he had already been hired, he started pre-deployment training and was able to quickly show he had probably spent too much time in his life drunk and bored. He was not a fit for a South Pole winter crew and was terminated. I made a point of talking to my crews about the importance of not posting stupid things on social media. I explained to them that it would be even better not to do stupid things, but if one does, posting them for the world to see forever is not a good idea. For my second crew, after experience gained with the first, I made it extremely clear that interviews were not over with their arrival at the South Pole. I explained that they had to get through summer to get to winter. That crew was extremely motivated and careful, and we did not lose a single member during the summer for any type of disciplinary reason.

While the selection process for winterovers was quite rigorous and continued during the pre-deployment emergency response training, teambuilding, and completing the summer prior to winter, there was a loophole. The loophole involved late hires who were filling positions opened by crewmembers who had voluntarily departed or were terminated in the summer. In these cases, there were no face-to face interviews, but rather, last-minute phone interviews. These phone interviews could almost take on an air of desperation, which we tried our best to conceal, as we needed to fill an important position. That ran totally against how we would have been when we were hiring months earlier. Making it worse, the candidate had not gone through emergency response or teambuilding with the crew. The phone interview and maybe a short time after their arrival at the South Pole was all we had to determine if someone was fit for a South Pole winter. By this time in my career, I had completed many phone interviews and realized that not physically seeing the candidate was a major handicap in determining their level of fitness for the job. In the case of a South Pole winterover, the need to determine fitness was much more than I had ever experienced. It was extremely difficult to determine by a phone conversation if a person who on paper was technically qualified would fit with an existing crew. The hiring of these people was my greatest staffing concern. In the worst case, it was possible to have them join the crew just prior to station close. There would be only a short amount of time to assess them and send them home if they were found unfit. The crew could then be stuck with them for the winter. During three South Pole winters, there were always a few late joiners on the crews, and either by masterful recruiting, successful phone interviews, or, more likely, just luck, on my crews there was never a person hired late who turned out to be a major problem.

During my first summer, I began to develop what would be my leadership style, which I held to for my nearly three years at the South Pole. The basic tenant was that I was "on duty" twenty-four hours a day and thus would not drink while on station. I wanted to always be ready for any type of emergency and considered a risk matrix with one axis for probability of occurrence and the other for the severity of the incident. In essence, you would probably have more minor emergencies, but you always needed to be ready for a big one. I never wanted there to be any question about my being fully fit for duty in an emergency and thus followed this regimen. At the same time, I wanted my crews to have the relief that alcohol in moderation provides and did not discourage drinking, but I did monitor it. I wanted to instill in them the need to exercise good judgment in their socializing and for them to take care of each other. One winter, on a Saturday evening during my walking rounds through the station, I noted a crewmember that surprised me with his high level of intoxication relatively early in the evening. He was seated in a station lounge with his friends and given his condition, conducting himself relatively well. A little later, I followed behind him and the two crewmembers, one on each shoulder, who were helping him navigate to his room. They got to the fellows' room, opened the door, and placed him in bed. The next day, I was approached by one of the three, who asked if they were in trouble for what had occurred the night before. I laughed and told him that in no way were they in trouble and that I was extremely proud of them for taking care of their drunken crewmember. The other thing I did religiously was walk thoroughly through the station, especially during Saturday nights, when people were socializing and drinking. I would visit the various social events, never staying long, and would usually engage in short conversations but in no way wanted to dampen any festivities. The crew really needed those times, and all I wanted them to know was that even if I was not engaged in many of the social activities, I was there if they needed me. There was nothing more important to me than being there for them.

During the first summer, I started a regimen of running and walking outside. My goal was to ensure I was outside every day I was at the South Pole. I began to log miles from the first day. I had done this with running in the past and had once gone over ten years without missing a day. In order to not miss days given my working around the world and frequent traveling, I had to do somewhat ridiculous things like running in the middle of the night outside hotels after arriving late at foreign locations. The first day I missed after more than a decade was in Africa, when I was staying at a place with many lions. After missing a day, I figured out a way to keep running, and it involved running with a local game tracker. He could at least spot the beasts, and he knew all he had to do was be faster than me to avoid danger. I kept calendars going back many years in which I had daily logged my running miles, and it was by this time over forty thousand. Having that kind of background, it was natural for me to keep track of my miles walking, and sometimes running, at the South Pole. What was more important to me than miles attained was to never miss a day. In two summers and three winters, I never missed a day. To do that, I was outside many times in conditions that humans, and I think most animals, would find unnerving and uncomfortable. My goal was not so much mileage as it was the hours and the experience of being outside. It was all about understanding and effectively dealing with the cold. The cold that had scared me during my arrival and would later, in the darkness of winter, drop to a level that made it almost unreal. Such cold as I was to experience in the dark of a South Pole winter would feel closer to being on an alien world than any earthly experience. An interesting facet to this walking was that I would always be alone. Prior to my arrival at the South Pole, I had always run alone and enjoyed most things alone. To be alone was important to me, and after the first couple thousand miles outside, I was as comfortable out on that ice as anywhere else, and probably happier.

Starting my walks at only a mile, I quickly worked up to more and added running as the weather warmed. I had run for many years, and while I had considerably slowed and suffered some chronic overuse injuries, I was still consistent. My running had gone from being a "runner" to a "jogger" and now was what I termed a "slogger." I was considerably slower than when I was younger but still consistent, and overall, I felt strong. For the South Pole, I had bought an incredible pair of running shoes designed for snow and ice, with cleats that would grip the ice. They had a zippered cover over the laces to keep out the ice crystals. These made running outside easier, and I was able to slog a few miles most days during the summers. I would then go for a walk sometime afterward. There were several routes I would take over the years, my favorite being to follow the remnants of the South Pole Overland Traverse, spot Road. I logged many miles on it.

The first summer I learned what to wear, what worked and what did not. Especially vexing was eye protection. I had no luck with goggles as they quickly fogged up. I tried the classic glacier glasses with the leather protection on the sides, but all I did was turn the lenses into minor glaciers. I ended up favoring no eye protection and simply covered my face with balaclavas and a fur hat to expose only slits for my eyes. I was able to do that even during the winters, in temperatures colder than minus 100°F, with no ill effects. On my second year, during the summer, to deal with the harsh rays of the twenty-four-hour-a-day sunlight that was reflected off the ice, I experimented with classic Inuit sun/eye protection. The Inuit are extremely ingenious people and have made their eye protection from natural products such as wood, bone, and leather for millennia. My sunglasses were a simple leather eye covering with small slits to look through that tied with leather straps around my head. I fell twice the first time I wore them as there was little peripheral vision and I would clip a boot on a hard edge of uneven ice as I walked. I pictured the Inuit wearing them, stumbling around the Arctic. I learned through practice how to walk while wearing them and how to watch for possible obstructions. They worked!

The icy surface at the South Pole is interesting. It is not what I knew growing up in the Midwest with snowfall. It is comprised of tiny granular ice crystals, and rarely are they the true geometric snow crystals that one experiences in the north. One question I asked prior to deployment was "is the surface slippery?" I was to experience the fact that, in general, it was not. At the South Pole, it never gets warm enough for it to thaw, and thus the ice crystals retain their shape. The ice crystals do not interlock and under normal South Pole conditions, will not form solid ice. It is never wet so does not form the slipperiest ice, such as what I experienced so frequently in the Alaskan Aleutian Islands. Compression by things like heavy machinery pulling equipment, such as the AARF (aqueous film forming foam) sleds that contain firefighting foam required as fire protection for the military aircraft, could create a slippery surface. After the ice was run over by the heavy sled there was an extremely hard, shiny surface resembling something left by a giant ski. This new, hard and level surface of compressed ice crystals looked inviting to walk on, but with a few grains of ice on top, it could be slippery. The natural form of the ice at the South Pole is sastrugi. Sastrugi is formed when wind blows over the ice in the same general direction over time. The sastrugi forms icy points in the direction of the prevailing wind. Walking on the virgin sastrugi can be challenging as you sink a few inches with each step and the surface can be quite uneven. Skis would certainly be preferred for long trips as you do not sink and can glide along the surface, saving much energy over walking. My first summer, I tried skis, as I had some basic training with the old cross-country version during U.S. Marine cold-weather training years prior. At that time, I had shown some aptitude with their use. After a couple of South Pole trials, I quickly saw that any aptitude I had for skiing was now gone. I then tried snowshoes and found them unwieldy and mostly unnecessary. I abandoned both. I was out for exercise and found walking, and sometimes running, was best for me.

I started my outdoor walking regimen the day after my first arrival, in October of 2016. It was an easy start, a mile walk wearing the issued USAP clothing. I had little experience in my new environment and took the situation quite seriously. I quickly found the Big Red parka heavy and somewhat cumbersome but used it until something I had ordered arrived. I also had a military surplus Generation III Extended Cold Weather jacket and pants set, which I had bought for seventy-five dollars. I had thought that, due to its light weight, I would use it for running. This military ECW proved much too warm for running but turned out to be a major bargain as I used it on many of my summer and winter walks. What I was awaiting via the mail was an anorak made in Norway of a heavy canvas material that was roughly what Roald Amundsen and his men wore while at the South Pole in 1911. Amundsen and his men wore furs at the start of their trip to the Pole but cached them as they were much too warm for the December summer temperatures on the polar plateau. They are wearing the heavy canvas anoraks in the famous photo of four of them looking at the tent they erected at the South Pole.

The day came when the Norwegian anorak arrived. I took it to my room, and upon first inspection, its raw beauty overwhelmed me. It was simply heavy, wax-coated canvas with a liner made of a type of flannel, and yet, to me, it had an aura better than any king's robe. The great Amundsen had worn such an anorak, and when I put it on for the first time, I felt that. Now I had a dilemma. My dilemma involved wearing such a historical looking item without it appearing as a sort of costume. It boldly stood out in the station's sea of issued Big Reds. I was concerned I was not worthy of wearing it. I decided not to wear it until I had been outside and had walked for over a hundred miles. The first time I would wear it would be the day Amundsen and his men had arrived at the South Pole, December 14. I wore it that day, and many after that, and found it extremely comfortable with its loose fit and ventilation. I found it was comfortable to about minus 55°F. It was not nearly warm enough for a South Pole winter, which was something even the great Amundsen had not experienced. It hangs in my home as a display now.

I slowly increased my walking miles and learned valuable lessons as I progressed. I experimented with clothing and face covering and had some minor frostbite while learning. The fact is that the South Pole during summer weather is not much colder than a cold winter day in the upper Midwest. Still, it could be lethal should one become disoriented and lost. For the most part, the weather did not change rapidly, and I made it a habit to check the local weather report prior to walking. I learned to spot bad weather approaching with high winds that might obscure visibility. Out a few miles from the station, I would continually scan back in the direction of the wind, as I could see ice clouds as they were coming my way. I learned that with a compass and some practice I could get back to the station in any condition, and that I should be aware of the direction of the winds, as I could use that for rough navigation. I learned not to think I could return to the station following my outgoing footsteps as the wind quickly filled them in with blowing ice crystals. A friend told me a horrifying story that had occurred some years prior. He was out about seven miles from the station on skis and lightly dressed when the wind picked up and obscured visibility. This fellow was quite a tough outdoorsman, seasoned South Pole veteran, and very adept on skis. Even so, he quickly found himself in a life-threatening situation. He struggled back but said, after that experience, he never went out like that again. I approached my walking, even though it was just for exercise, very seriously.

I learned to dislike walking on days when there was a phenomenon I called "flat light." During this very overcast condition, you could see things in the distance, but everything nearby was white and dull. When this occurred, you had no depth perception, and when you placed each step, you did not really see the surface. I would proceed carefully and quite slowly to try and avoid a fall, which could then result in an injury. On my second year, after I had the experience and clothing to get back in any condition, I had an interesting thing happen when I ran into the South Pole Overland Traverse, SPOT group in their special tractors bringing us fuel and coming from McMurdo. I was several miles out from the station and heading away from it when we met. Later, I heard that they were quite surprised to see someone in an Amundsen anorak solo, on foot, and that distance from the station. I walked farther and farther out, was extremely careful, and tried to learn to be ready for winter. People have crossed vast sections of Antarctica; thus a mere walk a few miles from the station should not seem to have any danger. But those making long trips do it in the summer; are self-contained with shelter, fuel, and food; and can make camp at any time. I had nothing and needed to be careful even during the relatively warm summers. Winter was entirely different, and the danger escalated with the unworldly low temperatures and darkness.

Boots were an especially important part of my South Pole outside wardrobe, and I started with the standard USAP issued boots made by Baffin and later changed to the old blue FDX Asics. I found the Asics to be extremely warm, and I never used the chemical heating packets favored by some. I also found the boots extremely heavy and clumsy and only wore them on the coldest days when I would get out some distance from the station. I purchased and tested several types of lighter boots and ended up favoring a lightweight mukluk made of canvas and moose hide. These mukluks were only rated to keep your feet warm to minus 40°F, but I found I could use them on days when the temperature dipped to colder than minus 100°F. This was only if I was in a situation in constant motion, such as on the flag line that ran out approximately a kilometer to the telescopes. I would do multiple laps on that. I would never wear the mukluks on extremely cold days when I went on my own track grid south. This was because if I experienced a problem and could not stay in motion, my feet would get cold. At worst case, cold enough to lose them. I was concerned that they would not be warm enough after a fall with an injury. In addition, they had a gum-rubber sole. After several miles in minus 100°F temperatures, I could feel the cold coming in from the soles and knew they were frozen. It caused me to wonder if extremely low temperatures could cause them to freeze and break. Having the soles freeze and break some distance from the station would be a very bad thing. I wore silk-based sock liners and heavy woolen socks. Keeping one's feet warm in such a lethal environment is extremely important, and I never had an issue.

While preparing for my second winter, I decided I wanted to try a new style of anorak based on Inuit and historical polar expeditions. I had seen an old photo of Roald Amundsen wearing wolf-skins (furs) that appealed to me, and I began a search to acquire my own. My search led me to a furrier in Russia that made custom clothing utilizing the fur of Siberian wolves. I contacted them, sent photos and my measurements, and after some back-and-forth correspondence on minor points, the anorak was completed. I made it truly clear that I wanted the anorak to be functional, sturdy, and not a fashion statement. I am not a fan of furs that are from animals raised for that purpose and find that process heinous. After some research, I found that the Siberian wolf was in a very overpopulated state and now so prolific that wolf packs were wreaking havoc on the Indigenous population's reindeer herds, killing pets, and even threatening children. These people's very existence and way of life was being threatened by the ever more aggressive wolf packs. With that, they were culling some of the wolves and selling the fur to bring in income. I could live with that.

I received the anorak, and as I had done with the Amundsen canvas one, took it to my room and opened the package. Although I had seen photographs of it from Russia, nothing prepared me for the first sight of it. It was shocking to note the beauty of the wolf fur and its weight and thickness. It had a silky material as a lining, which aided in putting it on and taking it off. I put it on by pulling it over my head, and as I had given the furrier my body measurements, it was a perfect fit. Initially it was almost overpowering to wear it, as it had such a wild look and feel. A few days later, I wore it for a summer walk and learned my first lesson: it was too warm for summer weather when one was really moving. After some experimentation, I found it needed to be colder than minus 50°F with no or little radiant sun to be comfortable walking in it at any speed. The rule changed somewhat if it was windy or if I was standing still; it could then be worn at warmer temperatures. The second lesson was that when I wore it, some people liked to pet me. I began to understand how a dog must feel.

I would wear the wolfskins on the coldest days and during the dark and cold of winter. I found them warmer and more comfortable than the modern extreme cold weather clothing I had. Minus 100°F and colder was not an issue when I wore that anorak, and I really felt the wolves' presence. The anorak, to me, had a wild, primordial feel to it that I respected and treasured. Those animals had died, and now they were protecting me from a frigid death. When out walking, I not only respected the wolves, but I also felt their presence and power. Wearing that anorak on one unforgettable cold, dark, and clear night, I walked under the light of a full moon. Some distance from the station, I stopped and stared long at it. I will admit to experiencing an urge to howl. I was then enveloped in a wave of sadness as I realized that those wolves had once stared at that same moon. I ask no pardon from anything earthly for wearing those furs but hope the great spirits of those wolves can forgive me.

One of my very few minor disappointments while at the South Pole was something I never expected to see, tourists. Most Antarctic tourists arrive via cruise ships and make landfall at the end of the Antarctic Peninsula. Some never even realize or maybe do not care that they have not even crossed or just barely crossed the Antarctic Circle. I imagine those cruise ships are fun and could be educational. I hope that they do all they can to prevent any kind of environmental damage that could be caused by their voyages. What I did not realize was that the South Pole had tourists, although in far fewer numbers than the cruise ships had. I had always endeavored to pay a price to immerse myself in a location I really wanted to visit and had several experiences where I paid a price of sorts. An example was a fascination I had with the South African Zulu War of 1879. I had read books on the subject and had a solid general knowledge of the war and its battlefields. Of primary interest were the great battles at Isandlwana and Rorke's Drift. I traveled to South Africa and, rather than an organized tour or even just a drive to the battlefields, decided to walk. It was late 1993 and the country was in turmoil with a great election looming. There was also an interesting practice occurring between taxis that ran from village to village, called the Taxi Wars, in which it became acceptable to shoot up any rivals and their passengers. It was probably not the best time to go. I tried to retrace the steps of the famed British Twenty-Fourth Regiment on its way to their great disaster. I wanted to see Zululand somewhat as they had and started my walk at Pietermaritzburg, where they had started. Of course, many things had changed since 1879, and I found myself at times walking on the shoulders of busy roads with cars hurtling past—the drivers probably wondering what the idiot with the backpack was doing. I walked every step, which turned out to be approximately 120 miles. Noting that, the men that ran the battlefields let me make camp at Rorke's Drift and Isandlwana. This was a rarity and an honor. It turned out to be a rather interesting experience at Isandlwana as the dead still reside there and, on that night, they tolerated me. After the long walk to get there, when I surveyed those battlefields, it was a very intense experience. I wrote nothing about it and claimed no silly first. It was just a walk. This was much preferred over arriving on a tour bus. At the South Pole, most of the tourists did just that, but instead of a bus, they arrived by aircraft.

The USAP does not actively support tourism at the South Pole but realizing that it is a fact, makes minor allowances for it. Most tourists arrive via a Basler aircraft charted by tour companies. Immediately after disembarking, they head straight to the ceremonial and geographic South Pole markers to take their hero photographs. Once sufficiently photographed, some groups receive South Pole Station tours on a not-to-interfere-with-station-operations basis. Station personnel lead the scripted tour and for the most part, enjoy the role. The tourists then head back to their aircraft and depart. The entire process is only a couple of hours and horrifically expensive. For the same price as that trip, a person could buy a sailboat and travel the world for quite some time. While we did not pay the price in such a gallant effort as to ski and dog sled from the Bay of Whales, as Roald Amundsen, or pull a sled from Cape Evans, as Captain Scott did many years earlier, my crews and I paid our price of admission in years of dedication to supporting important science projects.

When tour groups were allowed to visit the station there was never a shortage of personnel wanting to serve as guides. I noticed that the station personnel who served as guides were not always the most knowledgeable. The station's tour guides were supposed to follow a script that had been put together to show the station and its personnel in the best possible light. It was always important to show our station this way as one never knew who these rich tourists were and what they might say later. Knowing that, I was somewhat horrified when I overheard one of our very exuberant tour guides explaining in detail about all the alcohol that the station had and the inner workings of the 300 Club. The tour guide was quickly retrained and promised to follow the script on any future tours.

I was responsible for one faux pas when I was asked to say a few words to a group of visiting tourists. I would sometimes identify myself as the WSM while they were receiving their tour and discussed specifics that I knew they would like to hear about winter things, such as the low temperatures, length of darkness, size of the winter crew, etc. On this occasion I had not been sleeping well and was quite exhausted. I use that fact as an excuse for what happened. This group of around eight or so was quite unique as it included a couple that had been married after their arrival at the South Pole by a minister that was in their group. The expense of that wedding must have been staggering. My speech started normally enough as I identified myself as the wsm. Then I asked if they were the group that had a newlywed couple who had just been married. They answered yes. I then, for some reason, blurted out, "Well I've been married three times and can't really recommend it." I noticed the look of shock on the bride and groom's faces and then noticed my inspiring words being captured by the wedding's videographer for an eternal memory. Trying to recover, I stated, "Well, if it makes you feel any better, for me it got better every time." Realizing I had done enough damage and having no possible recovery, I quickly departed the group.

There was a tourist camp a half a mile or so from the station that was comprised of tents and had several vehicles. Some tourists who paid even more than the exorbitant fee charged to just fly in and out actually overnighted there. I visited the camp a time or two and said hello to the camp staff. I was amazed by what they had brought to camp to be able to feed and house their rich clients. They could only operate in the summer season. Just prior to winter, the tents were taken down and everything was packed and left in place for the next year. Winter storms would cover it all, and the next summer, it would be dug out and reerected. The tourists were usually quite wealthy, and from talking to one of their camp staff, some were very generous with tips done for mundane things. In general, I found the tourists to be nice people with most genuinely interested in Antarctica and our station, but personally, I would have been embarrassed to see the South Pole the way they did.

Others arrived on skis and even bicycles. I imagine someday they will arrive on modified pogo sticks and skateboards, because many want to claim some kind of "first." This will be done by creating ridiculous, obscure niches that no one else has already claimed. To me, these trips were simply safe, yet expensive, self-promotional stunts that created a circus atmosphere, which I found distaste-ful. I had little time for the antics of so-called expeditioners, with sponsors; aircraft to drop them off and pick them up; and satel-lite phones, traversing routes that had been done many times, all the while with expensive rescue plans ensuring their safety. Most of those trips are funded by donations, and an almost universal statement used by those seeking funding for their great adventure is that their objective is to "raise awareness about Antarctica." The truth is that their objective is to get someone else to pay for their trip. Many people are quite aware of Antarctica and its current chal-lenges. If someone was truly concerned about Antarctica and its fragility, I would imagine they would think twice about a trip that causes aircraft to fly unnecessarily and its other impacts.

While I believe those trips could be physically very rigorous, real danger was minimal, and there was absolutely no type of true exploration involved. I saw self-promotional attention seekers per-forming these stunts, with exaggerated publicity being of para-mount importance to the participant. It was just another example of the current "me, me, look at me" lifestyle. The product was then fed to an unknowing and gullible public via books, lectures, and other means of self-promotion. Some of the participants would later become "motivational speakers," who, I imagine, unmoti-vated people were supposed to look to to become motivated. Per-sonally, I do have an attraction to the idea of a longer ski trip, but I would consider it a very rigorous athletic event. I believe those ski trips would be extremely physically challenging and interesting, but I was horrified to see people who participated in them some-times getting international press that compared their very minor achievements to the likes of Roald Amundsen, Ernest Shackleton, or Captain Scott. Those brave men, who lived in the heroic age of Antarctic exploration, were true explorers, breaking new ground with their backs truly up against the wall. Their safe return was not guaranteed. They were playing for keeps. They are true heroes and there is no one like them today.

There were several such instances that were generated by the fact that up to now no one had ever crossed Antarctica "unsupported and unaided." What unsupported and unaided meant was subject to much interpretation. One of the best stunts started with a flight to an inland starting point, skiing over the same route that many others already had done and adding the USAP's South Pole Overland Traverse SPOT Road to complete a "crossing of Antarctica." The debacle ended on a glacier that had nothing to do with any early exploration and was hundreds of miles away from the ending points of the earlier historical expeditions at McMurdo, Cape Evans, Cape Royds, or the Bay of Whales. The SPOT Road is recut every year and used for tractors that pull fuel bladders and cargo from McMurdo to the South Pole. There are three round trips made in the summer. While it gets somewhat drifted over after storms, remnants of the road are always there in summer, and I favored it for exercise. The best part of our SPOT Road for the modern-day, self-proclaimed explorer is the fact that there are marker flags every several hundred feet and the entire road is surveyed for crevasses by ground penetrating radar. If only Amundsen, Scott, and Shackleton had been fortunate enough to have had such a well-marked, crevasse-free road.

The one type of summer traveler that I did have some interest in was the motorized expeditioner. There was no pretense of being a Scott or Amundsen among these folks. For them it was simply about getting from point A to point B, and in some ferociously fast times. Of interest to me was the fact that they were testing out new cold weather vehicular technology. The downside to this was the running of vehicles across Antarctica and the negative effects of that. During my second summer, Russians came through the South Pole in two incredible vehicles on their way to their Vostok station. It was an unusual route, and I was intrigued. They were filming the adventure and had a Russian "star" from their TV with them, who was unrecognizable to us. The producer was a beautiful woman of Armenian origin who was quite vivacious. As the

U.S. relations with Russia were somewhat sour at the time, they did not get an exceptionally warm welcome, but they did receive a standard tour of the station.

Over my life, I had developed a great respect for Russia and its history. I had no interest in their communist era, except for their incredible heroics during World War II. My interests in Russia went back earlier, to things such as George Kennan and his book *Tent Life in Siberia* written in the 1860s. Also, of major interest to me was anything to do with the USS *Jeanette* expedition of 1880. The Jeanette was frozen in and crushed by the ice while attempting to get to the North Pole. The ship and crew had followed a hypothetical warm water current called the Kuro Siwo, thought to exist by the German cartographer August Petermann. Utilizing this warm water current that supposedly flowed north through the Bering Sea, the expedition had planned to be the first to reach the North Pole. As it turned out, this warm current did not to exist, and the *Jeanette* was soon icebound and later, crushed. With the ship crushed, the crewmembers made a desperate journey in three open boats to try and get to land. In that journey, one boat and all crewmembers aboard it were lost. The two boats that made it to land ended up in a desolate and sparsely populated area in northern Siberia. The survivors then made an extremely difficult trek through the Lena River area of northern Siberia. That any survived at all was due to the incredible leadership and tenacity of several members of the *Jeanette*'s crew and the efforts of the local Indigenous inhabitants. The Russian czar at that time was extremely helpful with the rescue operation and the treatment of the survivors. I had read about recent plans to locate the *Jeanette* in Russian waters, and this did influence my desire to meet Russians.

I decided to do what I had never done with any of the summer tourists and make a visit to our small station store to buy them all South Pole hats, a pink sweatshirt for the producer, and a bottle of vodka. That bottle of vodka was the only alcohol I ever purchased while at the South Pole. I walked out to their two vehicles parked near the ceremonial pole. After waving at them through their windows, they welcomed me into one of the vehicles. Once inside, the vehicle's interior was quite impressive. Covering all the walls was a dark grey, felt-like material, which made it feel warmer. It contained a small kitchen area, and in the rear were sleeping quarters. There were five of us crowded into the vehicle sitting as best we could. This included the Russian TV star, the beautiful Armenian producer, a young and outgoing cinematographer, and a grizzled Siberian mechanic. I distributed my gifts, and they received them with much graciousness. The Russians then brought out bread and a type of pork-fat spread and some dried fish. Their English was not good, and I had no Russian, but we conversed as best we could. I could sense a slight tension in the air as I was an American and the countries' relations were strained. I am sure they wondered what I really thought of them and their country. I felt bad that we could not just meet as friends with no suspicions.

The vodka was opened, and my new friends insisted I have the first drink. I told them I wanted to but was on duty. To me, my duty was the year I was to be at the South Pole, and it would end with a drink with my crew once we were relieved by the incoming crew. This was a long way off, and understanding my situation, they made me some excellent tea instead. The conversation became less tense as the vodka bottle made the rounds, and they told me they had started with more than 100 liters of whiskey, but they had drunk it all. I was shocked by how fast the bottle of vodka disappeared between the four that were drinking it. At some point, they asked if I could get more. I told them I could not as we had a ration for alcohol. After some eating and drinking, I had some plastic trash in my hands and asked my new friends where I should put it. The beautiful Armenian producer was sitting on a trashcan with just a slight opening between her legs. She motioned to put it there, pointing down between her legs, and everyone, including her, exploded in laughter. I felt my face turning a bright red, and I muttered that I would just put the trash in my pocket. I was glad that I had extended myself and hoped the Russians enjoyed the experience. I wished them well on their way to Vostok.

We had our own USAP/ASC versions of tourists, and I learned to dislike the South Pole summers after experiencing the first one. For me, there were too many people blowing through with little to do and some outright boondoggles, where USAP/ASC and other government agencies had dreamt up some type of excuse to need to make the trip to the South Pole. Some of the worst were the U.S. military chaplains who would hold services to the pitiful hand-ful that wanted such. One chaplain, after arriving on an extremely late flight from McMurdo, showed her true priorities when all the arriving passengers on her flight were waiting together to start the mandatory station briefing. The briefing was a station requirement before these tired people could get their room assignments and go to bed. As people sat in the lounge awaiting the introductory video and discussion, she was nowhere to be found. I learned she was out at the Geographical South Pole obtaining her "hero" pho-tograph. The tired group was waiting, and it was quite easy to see what mattered most to her.

The summer in-briefs for arriving passengers were fun to do, and I did a lot of them. The arriving passengers ran the gamut in level of enthusiasm, from the starry-eyed people who had just ful-filled a lifetime dream of being at the South Pole to the veterans who had done it many times and it had long ago lost any sort of novelty. One memorable event occurred when a group had just arrived and one of the passengers tried to hand me his plastic pee bottle filled with urine. He did not know what to do with it and insisted I take it. I told him what the protocol for disposal was and he reluctantly placed it in his travel bag. The briefing started with a short discussion, which the station doctor or I would give, on the possible adverse health effects of arriving at the South Pole. The main concern was that a person had departed from sea level and three hours or so later was at an altitude of 9,300 feet. Making it worse was a physiological factor caused by the low barometric pressure that could make the body feel like it was exposed to an altitude of over 12,000 feet.

I would always start the briefing by asking how everyone was feeling. I would usually get a thumbs up or "great," especially from the excited personnel who had arrived at the South Pole for the first time. I would then tell the group that if they began feeling any symptoms of an altitude related illness, to immediately report to the clinic. The major symptoms involved headache, shortness of breath, and extreme fatigue. Most people arriving at the South Pole will feel tired, and they might also be dehydrated, with the dry environment and probably not drinking enough on the incoming flight. We explained to all incoming personnel that they might have trouble sleeping and that some people had been known to wake up gasping for breath because of the lower oxygen levels. We had a few that immediately exhibited signs that they were suffering from serious altitude sickness and were placed on flights back to McMurdo shortly after arriving. Some with milder cases were able to suffer through it with medical assistance from our clinic and felt better in a couple of days. This is a potentially lethal illness that requires immediate medical attention, with the best remedy being a quick trip down to a lower altitude. I never had an issue with the altitude but would feel a little tired on the day I arrived and maybe the next day too.

I continued the briefing with a video that explained the station and the arriving persons' responsibilities. These included things such as the fact that you were only supposed to take two, two-minute showers per week to save water. At the same time, you should constantly be drinking water to hydrate due to the dryness. There was also a short section instructing the newly arrived on the way to properly close the massive, freezer-like, exterior station doors so as not to disturb one's neighbors. At the end of the briefing, questions were answered and rooms assigned. The South Pole Station has a capacity for about 150 people. There were a few people who, due to the station being full, had the interesting experience of staying in one of the heated, Jamesway-Quonset-hut-style buildings a couple of hundred yards away. They had outdoor plumbing, and no one preferred those.

One of the first summer traditions is a small tribute for the U.S. Marine Corps birthday on November 10. The day after is Veterans Day. The USMC birthday is observed if there are Marine veterans on station that care enough to do something. I did, and every year there was a photo taken of any Marines on station with a Marine flag that I had signed by John Dale out on Wake Island. John Dale was a veteran of the great battle that occurred there in World War II. He had made a return visit years later, during which I had him sign my USMC flag. (I had taken that flag all over the world with me, including Iraq and Afghanistan.) We would also have a short ceremony in the galley, along with cake. This is a Marine Corps tradition. Veterans Day was slightly more formal as there were usually several veterans from various services on the summer crew. We would do a reading of a U.S. government proclamation honoring the day and take photos; if there was enough interest.

One year the ceremony was slightly marred when a National Oceanic Atmospheric Administration, NOAA, "officer" wanted to speak about the "uniformed services." This was her way of including her group of government employees who worked for the Department of Commerce and never served in the armed forces with the military. I did some research on this, and NOAA had put out a directive that specifically told employees that they were not veterans. Some of them seemed to have a hard time accepting that. Since that time, pseudo-veteran status has been granted their organization, much to my dismay. The fact is, they did not serve in the U.S. Armed Forces. They were civilians who could resign at any time, unlike military personnel. I would cringe when a NOAA "officer" would wear the uniform to a South Pole function as it looks remarkably like a U.S. Naval uniform but, to me, had the authenticity of a bartender on a civilian cruise ship. Even with my disdain for the mock-military uniform and veteran status for people who had never served in the U.S. Armed Forces, I must say that the organization does do great things, and all three of the NOAA uniformed people I had on my crews were fine crewmembers, and each contributed greatly.

Thanksgiving at the South Pole is a beautiful event. The galley is decorated and the staff go all out to prepare a meal described by some as the best they have ever had. Thanksgiving is a wonderful holiday to give thanks and to be at the South Pole, for many, a lifetime dream, which makes it all the better. It is a wonderful holiday as it can be observed by most any person on the planet, to give thanks for what they have. I gave a couple of Thanksgiving dinner speeches and liked to end with saying that if a person could think of no other reason to give thanks, to give thanks to being at the goddamned South Pole. Christmas is slightly different and observed in a mostly lukewarm manner as the program and most of its participants are not overly religious. One year we had problems getting volunteers to put out the Christmas tree and decorations. Other years we had more participation. The USAP is a government organization and thus does not promote religion, but it in no way discourages anyone's right to practice. As the program participants, in general, run to the left politically, Christmas at the South Pole takes on a very nonreligious nature for most.

News Year's is a major holiday at the South Pole, and there is a large party in the gym. The problem with that holiday is it will involve excessive drinking, and many of the station residents stay up extremely late. I always prepared my winter crew for the holiday with the fact that, historically, an idiot or two would end up being terminated for their New Year's Eve antics so to beware. There were several examples, but one that I think truly hit the trifecta for stupidity and bad judgment was the winter engineer who in his drunken stupor had described himself as a "human weapon." He then proceeded to try and convince others of that fact. Another epic meltdown occurred with a fellow who had arrived to work during the summer only. On the day of his arrival, I overheard him in the hallway talking to a coworker about his desire to winter. Interestingly, we were at that time looking for someone with his skill. But after just a minute of listening and observation, I was immediately against it. By this time in my life, I had developed a good ability to sense people with problems, and I quickly surmised that the fellow was full of them. I went to the summer Human Resources person and told her what I had heard and that I had no interest in having him on my winter crew. I then stated I was going to watch the guy as maybe I was wrong. I was not.

Soon afterward I began hearing somewhat bizarre stories of the fellow's after-hours behavior. One that caught my attention was how he had dealt with being turned down by several women for any form of companionship at a social event. He left the event and made his way to one of the lounges. Once there, he drunkenly requested another drunken individual to strike him in the head with a piece of wood. I will never know exactly why he wanted to be struck by the piece of wood but thankfully, the request was declined. To remedy the lack of being brained by the piece of wood, he started slamming his head into a wall. The positive in this was that, at that point, at least it was only his own head and not someone else's. While he should have been immediately terminated and sent home when queried afterward, his drunken friends declined to make any kind of formal written statements. Fortunately for the station, he guaranteed his release a week or so later when he thought it a good idea to take a snowmobile for a drunken ride. He did not kill himself or anyone else in the incident, but it sealed his fate. He was immediately sent home, and it was quite clear that he was not winter, or even summer, material.

The ice tunnels that run under the ice out from the main station draw some of the most interest at the South Pole. The purpose of the tunnels is to contain the freshwater piping that brings water from the Rodwells into the station for further treatment and distribution. These tunnels also contain the station's wastewater piping, which transmits station sewage out of the station. The waste is then injected into one of the old Rodwell bulbs beneath the ice where fresh water was once obtained but has been abandoned. The tunnels are cut through the ice; their total distance is around two thousand feet, including branches. Their dimensions are approximately six and a half feet wide and eight and a half feet tall. As the ice compresses, the tunnels' dimensions slightly diminish, and some summers, the walls and ceilings need to be trimmed with a chainsaw. It is quite an operation to behold, with a chainsaw operator doing the trimming and chunks of ice being pulled out of the tunnel on a sled. The tunnels run at a depth below the surface of about forty to fifty feet. An interesting feature is that they maintain a relatively constant, year-round temperature of around minus 50°F. While they may seem a cold place in the summer, they are relatively warm in winter. Walking down those lighted tunnels can be somewhat eerie when you ponder how many tons of ice are above you and then hear a D-8 bulldozer running aboveground. There are a few escape routes to the surface, in the event of a collapse, but if such an event occurred, it would be quite possible there would be no way out. I noticed while down there that there was a marked bowing of the ice from the massive compression. It was not a place that I really enjoyed spending time.

Some of the most interesting aspects of the ice tunnels are the makeshift "shrines" located in square cutouts along the wall for the first fifty feet or so. There are several objects enshrined there, such a Russian sturgeon and other items whose meaning is now lost to history. There is one though that I am very aware of and witnessed the event that caused its creation. During the summer before my first winter, we received a message that Buzz Aldrin, former astronaut and second man on the moon, was coming to the South Pole as a tourist. Normally the tourist flights are ignored by most of the station personnel, but once people found out Buzz was to visit the station, they became quite interested. On that day, our NOAA officer decided to wear his formal dress uniform for some reason. Other people who would normally be performing tasks in different parts of the station found their way inside and around where Buzz would be during a station tour.

The tourist Basler aircraft landed, and the passengers deplaned. A couple of station personnel who would be their tour guides met Buzz and his entourage at the aircraft. The tour group with our guides then began the usual walk to the station, several hundred yards away. I watched from a window as Buzz walked along with a fellow carrying a video camera filming every step. You would have thought he was back on the moon. Soon we received a frantic radio call from our tour guides that the tour group wanted transportation to the station. This was declined, as the USAP is quite clear that we have extremely limited responsibilities for tourists. The station does not provide transportation. I watched as the entourage continued to move slowly toward the station. Upon entry, it was learned that Buzz did not feel well. He was immediately seated in a chair near our first floor Designation Alpha, DA, entrance. Soon afterward, nearly everyone on station found a reason to be at that location. Our doctor was called, and Buzz's basic vitals were taken. It was apparent that he was ill. He was taken to the clinic where a further evaluation was performed.

At that point, there was a discussion among the tour group, which also included Buzz's son, who demanded that the group be allowed to overnight in the station. This was thought necessary because Buzz was ill, and the tour company's pilots were supposedly short on flying hours. Tourists overnighting in the station is not a usual event, and my boss, who was there for the summer, was in his office with the door closed on the phone with a NSF representative trying to get some direction. I was outside trying to deal with Buzz's irate son. I found the fellow rightfully concerned with the situation of his ill father, but we were working on getting transportation for Buzz. We also knew that the pilot for the tourist aircraft had enough flight time left to safely depart the South Pole. There was no need to consider an overnight in our station. I visited Buzz in the clinic and spent a few minutes talking to him as he noted my Explorers

Club hat. We were both members of that organization—he being an especially prominent, if not the most prominent, member. During my time in the clinic, I heard the end of a story Buzz was telling that had to do with Neil Armstrong. I found it fascinating that I was listening to a man who had participated in one of the greatest events in human history. He must have been perpetually pulled back to those days in 1969. Most people would probably expect him to constantly relive that incredible event and not care what he thought about the weather or other more mundane matters. After an hour or so, we received direction that we could send Buzz and his assistant to McMurdo Station on an incoming LC-130. The tour company would then depart on their own aircraft with the rest of their group. We transported Buzz out of the station to the awaiting LC-130, and it departed for McMurdo. There was discussion afterward by some, questioning how Buzz received special U.S. government evacuation while he was on a foreign-owned tourist company's trip. The aircraft Buzz departed on was a regularly scheduled USAF LC-130 that was already inbound when Buzz was found to be ill. The somewhat negative discussion surprised me as Buzz is a national icon of unparalleled status. Personally, I would have not had a problem with a special flight. At the same time, I was fine with the rest of the tourists on that trip making their own way out, and in the end, they did. The incident should have ended there, but for us, it did not, as it then gained South Pole immortality.

After Buzz departed, I noticed that the tissue he had used while he sat on the chair near our DA entrance was on the floor. I picked it up and, knowing its great value, presented it to two of my crewmembers who aspired to become astronauts. Several weeks later, I learned they had found a Plexiglas case for it and wanted to place it in the ice tunnel. On a Sunday, a small group of us met in the ice tunnel to commemorate and consecrate the shrine to Buzz. The shrine consisted of the encased tissue Buzz had used, placed on a plastic glass it was thought he had drunk from. We had an unveiling, which I started with a prayer of sorts. This was all on video and probably the only video of such an event for any of the shrines. It was our hope that this shrine remain appreciated and unmolested for the millennia it will take for that ice to flow into the Southern Ocean. It was also our hope that, if desired, Buzz could at some point in the distant future be cloned from any fluid preserved on the tissue. On the Plexiglas case that contained it, there was a warning of sorts: "Buzz Aldrin's tissue, medevac 2016. Caution: Space Germs."

During the winter, the crew designs and builds the South Pole geographical marker for the following year. In the summer, on January 1, the marker is unveiled and planted in the ice at the Geographical South Pole. It is a popular tradition, and many station residents show up for it. The area manager from the Denver office carries the new geographical marker, which is on a pole and under a cloth cover, out to the Geographical South Pole area. As the ice moves at around thirty-three feet per year, the new geographical marker and the Geographical South Pole sign will need to be placed in a new location, which is set up prior to the event. The area manager removes the cover, unveiling the marker to the assembled group who will marvel at its beauty. The pole marker is then passed along a line of people who each pass it on to the person standing next to them. This is a great thing and makes everyone feel a part of the tradition. Eventually it ends up at the geographical pole and is planted in a precut hole in the ice. I learned that in the past it was simply planted by whoever was the last person in line. That person may have been someone who had spent a great deal of time at the South Pole or just someone visiting for a few days. During my first summer, I planted the pole marker. Later I tried to establish a tradition that the incoming WSM would always be the one with the singular duty to plant it. I felt it important that someone who had interacted with and relieved the outgoing winter crew would honor them by planting it. When I planted the pole marker for 2017, which had been made by the 2016 winter crew, I said under my breath, "I now declare this icecap Hvitt Hjem," which is Norwegian for "White Home." I claimed the South Pole for myself.

THE WINTER VOYAGE

During each of my three winters, my crews and I anxiously awaited the day the station closed and winter officially began. Station closing means that the station will no longer receive aircraft due to the coming darkness and low temperatures. This, in essence, seals the winterover crews in with no way to leave the South Pole until the beginning of November. February 15 is the target date, but that can fluctuate slightly due to weather problems or aircraft availability. Studying past meteorological data has shown that around that date, it starts to get too cold for the LC-130s to fly. Most of the remaining summer people are extremely excited to leave and have made plans for what they will do in Christchurch and other travels after departing the ice. The winterovers are even more excited about the summer people's departure. It is the start of winter, and the winterovers will now own the station. This transition of people can be a little clumsy at times as the winterovers feign sadness for the summer people's departure and the summer people know it is somewhat insincere. This is readily apparent as the summer people begin the walk to their awaiting aircraft while the joyful winterovers shout goodbyes.

My first winter crew had a memorable day when the last aircraft carrying what remained of the summer crew departed. This would be the first winter for most of us, and this was the day we had long awaited. Winterover crewmembers were receiving the last cargo and fuel from the aircraft and serving as the emergency fire force. This is required for all LC-130s for landings and departures. After all the summer crew had boarded the aircraft and the door shut, the crewmembers that would remain for the winter were furiously snapping photographs of what would be the season's last LC-130. A group of us watched from the station's second floor observation deck. It was a euphoric moment as the LC-130 revved its engines, taxied, and departed. We were ecstatic as the aircraft flew a lazy circle, came back toward the station at an extremely low altitude, and gave our winter crew the wing wave. The wing wave is done by the pilot, who rotates the aircraft slightly in each direction to symbolize a goodbye. Cheers went up! It was now winter, and the forty-six of us would have no way to leave for around nine months. The next two winters started relatively the same, but with smaller crews of forty-two. For whatever reasons, the LC-130 pilots chose not to do the wing waves those two years, but the euphoria was the same. I have heard that some crewmembers are greatly affected by the departure of that last flight and find it unnerving. I have never really seen that in crewmembers, although they might privately feel that way. What I saw was the same joy that I felt. This was the moment I had waited for, and I believe my crewmembers felt that way. It was the true start to why we were there. Much awaited us.

I started my winters with the calling of a quick "All-Hands Meeting." The main purpose of the meeting was to steady the crew and make winter room assignments. During the winter, the crew size is relatively small and there is much more room in the station. Crewmembers usually prefer to move into the slightly larger "A" wing rooms, which are also closer to the galley. Spare rooms to hold extra gear are made available to provide as much comfort as possible to the winterovers during their long stay. I stayed in the same room I was to have for all three winters, B-219, due to its proximity to the B-1 lounge and its end of the hallway closeness to a station exterior door. I was to exit that door many times. The B-1 lounge had been problematic during some years, but not when I was there. While I had heard stories of somewhat wild happenings occurring in that lounge, I found the most raucous evenings sedate compared to other places I had been. After seeing teeth and blood on the floor of the dayroom after a vicious fight on Shemya Island, Alaska, the B-1 lounge seemed quite tame. Still, I wanted to ensure things did not get out of hand; over three winters, they never did. During all my winters, but more so the second and third, most of the crew lived in the preferred A wing. There had been a comment made at the end of the second winter that I had not allowed someone who wanted to be in A wing the opportunity. Hearing that, for my third winter, the B wing had very few residents. I had my room at the end of an empty hallway. At some point during that winter, the flores-cent lighting in the hallway failed and the only light was from one unit at the end of the hallway, near my room, that eerily flickered. Looking down that hallway, it looked like something out of a mur-der movie. I preferred it stay that way and told our electrician not to fix it. I enjoyed the walk down that foreboding hallway to enter my room, which was wonderful. With the great ship-cabin config-uration, tattered flags battered by Antarctic winds on the wall, my hanging polar outfit, and my books, it was a sanctuary.

During the week or so after the station closes, Basler and Twin Otter aircraft from a Canadian company that has the small aircraft USAP contract, make transits through the South Pole Station on their way home. Some of the transits with the Baslers only require fueling, and then they fly on to the British Rothera Station near the coast. The Twin Otters, due to their more limited range, need to overnight, and that means a few strangers on station, to the mild dismay of the winterovers. While never overt, at least some of the winterovers silently wish them to leave, and the sooner the better. After my first year, I noted this. Not wanting these flight crews to feel totally unwelcome, I endeavored to be more involved, friendly, and outgoing with them. I wanted them to feel like guests in our station and to leave with good memories of our hospitality. Even with that, I was always happy when the last small aircraft departed.

I would joke with my crews that we would be checking that aircraft for stowaways, as it was truly the last way out. After the last departure, we would remove the airfield markers that lined each side of our skiway and dismantle the aircraft fueling system. The aircraft fueling system would be stored for winter in a different location, and the skiway would cease to be groomed by heavy equipment and would return to its natural state of sastrugi until the sun rose and we prepared it for our opening flights. Once completed, the station was now on its own, and we were completely physically cut off from the world.

A movie is what really starts a South Pole winter. It is actually the showing of three movies. The tradition is the showing of the three versions of the movie *The Thing*. This usually happens during the first weekend after the summer people have departed. The gym is set up as a theater, with couches and chairs borrowed from lounges; popcorn made; and the fun begins. *The Thing* is the natural movie of choice to show at the start of a South Pole winter as all three versions are about crews at Arctic and Antarctic stations facing a lethal and upset alien that they removed from the ice after its spacecraft crashed. Mayhem and death then ensue. I would begin with a short winter welcome speech and then discuss things related to the movie. I liked to ask who believed there was a "station gun?" The station gun is a myth of sorts that somewhere on station is a gun to which only the WSM has access. It is to be used in a time of great emergency. Most crewmembers do not believe the myth, but there is no great conviction with that. After my query about the station gun, I would ask, "How many bullets are there?" I would listen to the various answers then state that the number was the same as the crew size, minus one. This would usually get a laugh.

I watched all three versions of *The Thing* each winter and really enjoyed that first real social event with my crews. You could feel the excitement in the air during movie preparation. During my second winter, an event occurred that was troubling. On the first Saturday night after the summer people had left, we scheduled the movies. We were setting up the gym when the dilemma unfolded. We had a Canadian, transiting flight crew staying overnight in the station, and I was approached by several of my crewmembers asking if the Canadians would be allowed to participate. They made it quite clear that they did not want them to be included. I then asked several additional crewmembers and came up with a strong majority "no." They wanted to spend the evening with only their crew. I understood and respected that. Trying to avoid an international incident, I posted a sign outside the gym and called the event a "meeting." In case any of the Canadians wanted to watch the movies, I then set up all three versions of the movie in a lounge with a large TV. About an hour prior to the start of the "meeting," one of the Canadians wandered down to the gym. I spoke to him and said the event was for the winterover crew only, but I had set up a movie room for any visitors that wanted to view the movies. Oddly, he looked at me as if I were crazy and stated, "That's weird." He exited the gym in a somewhat disturbed shuffle. His reaction made my blood boil, but I calmly stated, "It's weird that we spend a year of our life at the South Pole." I was surprised that the Canadian acted the way he did and figured he did not really understand that this was more than a movie to most of the crew. I thought of how I would have been should I have been in a similar situation with something like a military unit holding a function solely for their people. I would have exited gracefully and wished them luck. He didn't.

This marred the evening slightly for me as I did not want the situation to escalate, and I wanted the crew to have a great experience. This would be a once-in-a-lifetime event for most of the crew. Not everyone agreed with the omission of the Canadians. To make matters more volatile, one of the over-exuberant winterovers had drawn a Canadian maple leaf with a red circle around it and line through it on the meeting notice outside the gym. That was exactly the type of thing I didn't want. I thought it was a rude gesture toward the visitors that should not have been done. I received a very heated email from one of my crewmembers who was friends with the Canadian who had walked into the gym, and she was quite irate. I spoke to her the next day and, I think, had a meaningful discussion. To her it was simply a movie, so why exclude anyone. To me and most of the crew, it was a rite of passage, and simple spectators were not welcome. While she did not know it, the crewmember who sent me the angry email was special to me. If I had a daughter, I would want her to be like that crewmember. After receiving the angry email, I, who have never been a father, had a taste of what must come with that territory. While I did not feel good about the decision I had made, I believe it to have been the best possible. Saying this, I was quite happy when on my third winter there were no transiting Canadians on station the night of the viewing.

The next traditional movie occurs at midwinter, in June, with the showing of Stanley Kubrick's *The Shining*. This move is shown as its plot revolves around a family who are caretakers at an eerie Colorado hotel during winter. There is certainly a similarity with the winter experience at the South Pole Station as the family is also cut off from the outside world by the ice and snow. In the movie, mayhem and murder ensue, and the eerie scenes that Kubrick created are not soon forgotten. It is another rite of passage for a winter crew. And while entertaining, it is spot on that we are, for the most part, cut off from the outside world. One of the aspects of my job was to ensure that murder and mayhem did not ensue.

I had learned from previous remote projects that conducting scheduled, weekly recreational activities was a good thing. I had tried movie nights and special programs at other locations, with mixed success. I had found it hard to compete with what most people on remote sites, especially when they work a six-day week, want to do on their only night off, drink. At the South Pole, things were different as the people were different. As one who respects history and adventure in general, during my first year, I started "Adventure Movie Night with Wayne," AMNWW. This occurred in one of the lounges and started at 7 p.m. on each Saturday night. My intent was to provide an alternative to early drinking, and the event would end with plenty of time for participants to still engage in some form of debauchery afterward.

For three winters, I started AMNWW with Antarctic-themed movies, beginning with the classic *Scott of the Antarctic*. This beautiful movie, made in 1946, is a testament to an age when Captain Scott was recognized as a heroic figure overcome by natural forces and dying with his comrades as true English gentlemen. This is my view, which clashes with some of the more contemporary Monday-morning quarterbacks who have written books on the subject. These authors have little experience in such matters and have written books critical of him and his expedition from the safety of their desk. Due to these books, many people with no real leadership experience in the Antarctic or, in most cases, anywhere else, parrot the opinion that he was a poor leader. I hated that and wanted to present a more balanced view on all the polar greats. They all had flaws, as we all do, but they all were great men who had accomplished great things. AMNWW was not just a movie night as there were also short discussions. While I do not consider myself an expert on polar exploration, I have a sound background that could answer most questions, and I usually had informed crewmembers who could add to the discussion. During my last winter, I added to the AMNWW experience by bringing items from my personal collection, which I would discuss and let the attendees hold. Examples were shards of wood that were from the spar of Shackleton's ship the *Endurance* and a candle holder from Scott's ship the *Terra Nova*. I really wanted to provide a sense of history to receptive crewmembers. This would make their winters more interesting experiences and would be a tribute to those that had gone before us.

Over the next two months after the station closes, the sun gets lower and lower on the horizon and around the third week of March enters a period known as civil twilight. During this time, one can still see outside, but it becomes increasingly dimmer. While winter officially starts around February 15, with the departure of the summer people, for me, the real winter did not start until it became dark. Darkness at the South Pole is a gradual process, with the sun getting a little lower on the horizon each day and going through three periods of twilight. A rite of passage for a winter crew comes around the first week of April, just before total darkness, when the station window covers are installed. The window covers are pre-cut cardboard installed to prevent station lighting from escaping, as the light can affect the aurora cameras that are recording on the roof. After the windows are covered, the station becomes a different experience for the crew. The station then has more the feel of a spaceship or space station, with the crew safely inside and an unseen lethal environment on the other side of the walls. Most crew-members enjoy the covering, and by the time the window covers are installed, you cannot really see anything other than your own reflection in the heavily tinted windows anyway. Some months later, the crew will enjoy even more the removal of the coverings and the return of the light.

The station's outside lights and the personal headlamps used in the darkness are all required to be red, as that will least affect the astronomical projects that seek light from distant galaxies. The lights cast their reddish glow on the nearby ice, which further helps create an alien planet feel. Darkness at the South Pole and the return of the light some months later are gradual processes. Around the first week of April, nautical twilight begins, and by the end of that, it is somewhat dark, but with a distinct glow on the horizon. Astronomical twilight follows, which is nearly dark, and then, on the second week of May, it becomes truly dark. Darkness at the South Pole is almost five months in length, and this period was the greatest attraction for most winterovers. The South Pole does not have the beautiful scenery that McMurdo and Palmer stations have, with their mountains, volcanoes, and a freezing and unfreezing ocean. Also, the South Pole Station does not have their wildlife, no penguins, seals, whales, or anything else. What the South Pole does have is incredible winter night skies and magnificent auroras. The auroras are caused by the energetically, electrically charged particles that accelerate through magnetic fields in the upper atmosphere. I have no comprehensive understanding of that process but can say that the shimmering, mostly green lights that appear in that dark night sky are certainly beautiful. During periods of high intensity, they would cast an eerie green light on the ice as I walked.

I would prepare my crews for the cold and darkness by conducting an outside travel safety briefing. This would occur during a scheduled All-Hands Meeting as we neared total darkness. These presentations became more personal and more informative during my second and third winters as I was able to use information gathered from my ever-increasing time and mileage outside. I would start by stating, "There is death outside that window." With what I had learned, I do not think I was being melodramatic. The environment outside the station, especially in winter, is potentially lethal, and I wanted crewmembers to respect that fact and take all outside travel seriously. I was most concerned about the crewmembers who rarely left the station and may not have fully understood the hazards when they did have to go outside. During a winter at the South Pole Station, with its heating system, pleasant lighting, comfortable berthing, and events such as "steak night," it is possible to feel somewhat insulated from that frigid and quite lethal environment on the other side of the window. I did not want that to happen. I discussed first how to stay warm with proper clothing selection and how to wear it. Next came how to travel and then, what to do in the event of an emergency. I had heard stories of crewmembers on other crews in past years who had become disoriented while outside. The stories, to date, all had positive outcomes, but I felt some of those crewmembers were simply lucky. The main thing I stressed was to dress warm enough for any temperature. It was also important to have someone know you were going outside, where you were going, and when to expect you to return. In addition, crewmembers should carry fully charged radios and know how to use them. We had a missing person procedure, and if we had a situation occur with an outside disoriented crewmember, we would immediately remove all the interior window coverings and turn on the station's roof-mounted emergency beacon. Fortunately, we never had to do that, and most people became quite comfortable and adept at outside travel.

About midway through my second winter, a WSM who should have been my relief in November abruptly changed his mind on the day he was to fly to the Denver office to begin his assignment. This put the program in a bind to fill the position and was quite a surprise, as the fellow seemed extremely interested and had formally accepted the position months earlier. I told Bill, the area manager, I was willing to help any way I could. A plan was developed that I would do a back-to-back winter, leaving around mid-November and returning mid-January for the next winter. For me there were not many positive aspects to doing this, but the major enticement was that I would complete a third winter as the WSM. This was something no WSM had ever done before. In the sixty-four years the South Pole Station has existed, only two other WSMs had completed two winters. I would be the first to do three, and the first to do a back-to-back winter in that role. This was not of any historical significance as the record at the South Pole for winters was around fifteen. One of our scientists achieved that remarkable number, with another following closely behind at fourteen. Not to take anything away from their great achievements, but it needs to be stated that their duties are extremely specific, with workspaces not in the elevated station. Other than required community projects, they made their own schedule. For the right person, multiple winters in such a role could become almost normal. The WSM, who is responsible for nearly everything, is a completely different experience, hence the vast majority did a single winter. There is a reason for that, and I know several past WSMs who had rough winters I am sure they wish to forget.

I wanted something else from that third winter, and it had to do with the Antarctica Service Medal. The Department of Defense issues the Antarctica Service Medal after someone has spent a mere ten days south of sixty degrees latitude. It is highly coveted by the military, due to its rarity. Through the years, the time requirement to be eligible for the medal has been shortened. This was done to ensure military members on short deployments qualified to receive it. There is a story that a general who wanted one shortened the criteria to its current ten-day requirement so he would get one. Unlike most military medals, civilians can qualify to receive it. Civilians working for the USAP receive the medal for their time at the South Pole, McMurdo Station, Palmer Station, field camps, and ships. For the first year, a South Pole winterover receives the medal with a bronze winterover clasp. This metal clasp goes on the ribbon part of the medal. For the second year, it is a gold winterover clasp, and for the third, the clasp is silver. The silver is the last and the highest. There is nothing higher after the silver clasp has been awarded, even if one completes another winter. I wanted the silver winterover clasp that is awarded after three winters as the WSM. I would make extremely minor history by achieving that, which meant that one of my real attractions for doing a third winter at the South Pole was a small sliver of silver.

FOOD

Food, specifically good food, is one of the most prominent desires of crewmembers during a South Pole winter. One challenge is how one defines good food. I believe that most of the galley staff at the South Pole are proud of the food products that they produce. I think it amazing that they can come up with the food dishes that they do as almost everything has been frozen. As a Marine with expedition experience, I consider most food to be good food. Unfortunately, my definition was not really the rest of the South Pole crew's definition. During my first year, the galley staff followed set menus provided and coordinated by their specific company. For the most part, the meals were okay, but just okay, and complaints about the food started coming in early in the winter. I remember eating ribs that were as hard as jerky, but as I like jerky, I did not really mind. The crew thought otherwise. During dinner I once noted that hanging over the roast beef on the serving line was a hastily made sign scrawled with a marker that said: "Sorry I burnt it." While I appreciated the cook's honesty, one look at the charred remnants was really all it took to know the beef was burnt. One evening I walked into the galley for dinner and noticed the tables were quiet. I looked at that evening's meal on the serving line and saw a ghastly looking white fish with a pink sauce. I stared at it, and since I have never been known to have a poker face, my expression—and maybe my muttering while viewing the fish—was all it took to start the tables in rounds of laughter. I had no fish that night and neither did most of the crew.

During the start of that winter there was a major project to clean the ventilation ducts in the galley. Once the job started, there was no way to cook in the galley for over three weeks. With the galley unavailable due to the duct cleaning operation, we set up several large industrial-size microwaves and cooked what we were told were "gourmet" microwave meals. USAP SPOT personnel, while driving on the traverse to the South Pole, had eaten these meals, and according to them, they were much enjoyed. I well remember the first of those meals, which was supposed to be meatballs. After cooking, the meatballs were as hard as cue balls. Attempting to cut them could bend a knife. As it was a first experience, and thus a novelty, the crew was quite amused. The amusement stopped shortly after, with a series of maladies that came along with each new "gourmet" dish. This went on for weeks. Ramen noodles and peanut butter became staples as people began to avoid the microwave meals. After the duct cleaning was finished, the galley was placed back in operation, and for a while, people were quite happy. That did not last long. In that galley crew's defense, they did cook some outstanding meals, but I do believe they were constrained by their off-ice superiors to follow a very rigid menu that did not really showcase the culinary talents they had. I was told that one of the people that was doing the company's overall menu planning had experience working in homes for the aged. That would explain the fish and pea stew that someone with no teeth might have enjoyed.

Among some crews there is an unofficial tradition to rank the crew as far as who would be eaten first in the direst of emergencies. Making the top of the list could happen for several reasons. First, because the crewmember was not a very pleasant person. In that case, it might be wise to finish him or her off first as it would be better for the crew overall. The second reason for making the top of the list might be body type. Someone heavier or overweight could be a much better food source than a skinny person. I had several on my personal list, one for being an extremely funny guy and a joker. He was very funny to a point, but with much repetition over the long winter, not as funny. I enjoyed telling him the fate of a Mr. Collins from the *Jeanette* Expedition of 1880. Mr. Collins liked to pun, and the crew found him extremely funny for about an hour. He then became insufferable. His bones were found in the Siberian Lena Delta after the expedition collapsed, and there were theories that cannibalization may have occurred with at least one of the dead crewmembers. While I did not personally believe that it had happened, I wanted my joker to know the story well. He was a fantastic fellow, and I was glad we never had to eat him. I had another crewmember who I really enjoyed, and he was on the heavy side. Not obese, but well rounded. I told him that he had worked his way to the top of my list. I remember seeing him once in the galley at a mealtime and the subject of my own who-to-eat-first list came up. He was having desert and I told him to go ahead and have another.

There are three important formal dinners during a South Pole winter. The first is the Sunset Dinner that occurs on around March 21 and celebrates the setting of the sun. This is the first time that a new crew will have a formal celebration together. The second is the Midwinter Dinner around June 21. This dinner celebrates the middle of winter and the sun being at its lowest point below the horizon. The last is the Sunrise Dinner celebrated on September 21, the autumnal equinox, that celebrates the return of the sun. Of the three dinners, the Midwinter was by far my favorite as it is a historic Antarctic tradition. Expeditions from the heroic age of exploration went all out on this occasion, and there are great photographs of them in their winter quarters around very decorated tables. During my second winter, I had historic expedition sledging food from that era made up as pre-dinner appetizers. This included actual pemmican from Canada I had brought, which is a mixture of lean beef with some fruit added. I had the galley staff make up sledging biscuits, using a recipe we thought was close to what early

Antarctic explorers had used. Then a "hoosh" was made, which is a combination of the pemmican mixed with sledging biscuits and water. I had saved the dried fish that the Russians, who I had met in the summer on their way to Vostok, had given me and set it out. Many years prior, Roald Amundsen on his way to the South Pole had brought dried fish, mostly for his dog's food but also for use as trail markers. My hardier crewmembers readily ate the fish. The crew good-naturedly sampled the appetizers but could never have appreciated the way men who had pulled sledges for hundreds of miles had felt about their food. When you read the diaries of the early Antarctic explorer's, food was probably the thing most on their minds. Sometimes, when the galley served a soup that I found exceptional, I would make a remark to crewmembers about our inability to imagine the relish with which those early explorers would have eaten such a soup, if only they could have.

Food is probably the number one morale factor during a South Pole winter. During my first winter, I ate in the galley with the crew but at a separate table. At that table regularly sat several other crewmembers who liked to read. It was a quiet table, and we sat apart, with plenty of space between us. From the next table, I would hear bits of amusing conversations the crew would be having. I would note that during times with more outrageous discussions, several crewmembers would glance my way to see my reaction. I tried to show no emotion, but with some of the discussions, it was difficult as I found it so funny. During one dinner, I overheard a table conversation that was quite crude. It was foul enough that I was getting ready to intervene as a young lady was seated among a group of men. I did not do that as she jumped in with her own contribution that really upped the level of disgustingness. During my second year, I started eating alone in my office, which allowed me to eat in total privacy and quiet and read. Around midwinter, during a discussion with a crewmember in my office, he asked if he could ask me something personal. I said of course. He asked if the reason I did not eat with the crew was some type of leadership thing.

SOUTH POLE WINTER 2017

1: Wayne White
2: Les Lemon
3: Matt Krahn
4: Doug Howe
5: Bill Johnson
6: Viktor Bajbiklow
7: Andrew Nadolski
8: Zach Morgan
9: Mike Rice

10: Peter Gougeon
11: James "JP" McMichael
12: James Casey
13: Peter Bammes
14: Martin Wolf
15: Stephen Ashton
16: Josh Grivy
17: Mike Pintur
18: Grantland Hall

19: Daniel Michalik
20: Jason Spann
21: Eileen Sheehan
22: Hunter Davis
23: Eric Hansen
24: Garon Jones
25: Josh Neff
26: Catherine Dudley
27: David Riebel

28: Brett Baddorf
29: Sarah Baddorf
30: Tyler Butler
31: Stephanie Olcott
32: Gavin Reynolds
33: Brian Vacha
34: Rick Osburn
35: Jerry Everhart
36: John Dinovo

37: Matthew Smith
38: Clint Perrone
39: Adam West
40: Jeff Keller
41: Ryan Clifford
42: Gavin Chensue
43: Jack Clifford
44: Zach Kinberg
45: Robert Schwarz

Not Pictured: Kim Williams

FIG. 1. The 2017 South Pole winter crew. Photograph by Hunter Davis.

FIG. 2. The 2020 South Pole winter crew. Photograph by Yuya Makino.

FIG. 3. Author's station sign-out board the day he attained four thousand miles outside. Photograph by the author.

FIG. 4. Aurora over science projects. Photograph by Daniel Hampton.

FIG. 5. "Chinese" Basler departing the South Pole.
Photograph by Daniel Hampton.

FIG. 6. Author back in the station after a walk. Photograph by the author.

FIG. 7. First flight into the South Pole at the end of winter. The author is pictured walking with his boss, Bill. Photograph courtesy of the author.

FIG. 8. Ceremonial South Pole under the aurora. Photograph by Daniel Hampton.

FIG. 9. Captain Scott's Discovery Hut at McMurdo Station.
Photograph by the author.

FIG. 10. South Pole station greenhouse, a place the author spent little time.
Photograph by the author.

FIG. 11. LC-130 arriving at the South Pole. Photograph courtesy of the author.

FIG. 12. The 2020 midwinter dinner crew. Photograph by Yuya Makino.

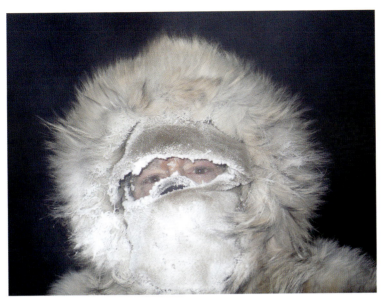

FIG. 13. Author walking in minus 88 degrees Fahrenheit under a full moon. Photograph by the author.

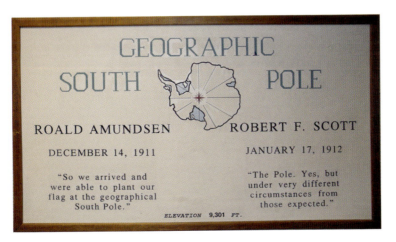

FIG. 14. Old Geographic South Pole sign, now on display inside the station. Photograph by the author.

FIG. 15. The end of winter and return of the light show drifting
behind the station. Photograph by the author.

FIG. 16. Author out running, as always, alone. Photograph by the South Pole summer fire department.

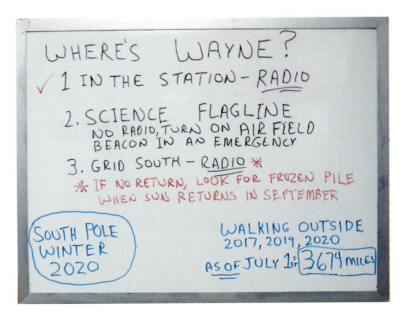

WHERE'S WAYNE?

✓ 1 IN THE STATION – RADIO

2. SCIENCE FLAGLINE
NO RADIO, TURN ON AIR FIELD
BEACON IN AN EMERGENCY

3. GRID SOUTH – RADIO ✳

✳ IF NO RETURN, LOOK FOR FROZEN PILE
WHEN SUN RETURNS IN SEPTEMBER

SOUTH POLE
WINTER
2020

WALKING OUTSIDE
2017, 2019, 2020
AS OF JULY 1ˢᵗ 3674 MILES

FIG. 17. Author's sign-out board with instructions in the event of no return. Photograph by the author.

FIG. 18. South Pole station panorama. Photograph by Daniel Hampton.

FIG. 19. Geographic South Pole sunrise. Photograph by Daniel Hampton.

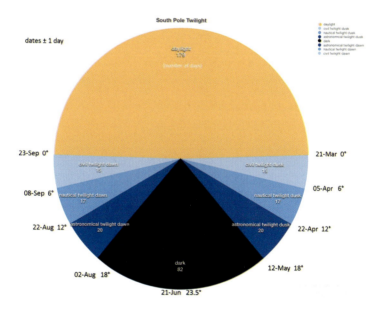

FIG. 20. South Pole daylight, twilight, and darkness chart.
Chart by Robert Schwarz.

FIG. 21. Dr. Geoffrey Chen outside under the spectacular aurora.
Photograph by Dr. Geoffrey Chen.

FIG. 22. South Pole telescope at sunrise. Photograph by Daniel Hampton.

Summer view of the storage area at the back of the
South Pole station. Photograph by the author.

FIG. 24. South Pole sunset while out walking. Photograph by the author.

FIG. 25. Self-portrait during sunset while out walking. Photograph by the author.

FIG. 26. This cross is a tribute to Able Seaman Vince of the Discovery Expedition. He died in a storm in 1902 near Hut Point. Photograph by the author.

FIG. 27. Having fun while out walking in minus 98 degrees Fahrenheit, a mile and a half from the station. Photograph by the author.

FIG. 28. Winter walking in darkness and a warmer temperature. This is indicated by the lack of ice on the author's face. Photograph by the author.

FIG. 29. Siberian wolfskins that the author used on the coldest days. He always felt the spirit of those magnificent animals. Photograph by the author.

I laughed and told him no, it was just that I liked to read. After he left, I gave the subject more thought. There were also timing reasons I ate alone: breakfast was just a grab and go, and during lunch, I might be in the gym. I also had walks to do outside that could be at a mealtime. But all in all, and if I wanted to be completely honest with myself, I had to face the fact that I did prefer being alone while eating. It was getting so that being locked away at the South Pole with a crew of forty-two was not alone enough.

At the next All-Hands Meeting, I told the crew that one of them had asked me why I did not eat with them and if it was a leadership thing. I let the crew know, in a humorous manner, that it was not a leadership thing; it was because I liked to read in my office while I ate. I then stated I would eat with anyone that wanted me to eat with them, and if they wanted, they could have a "goddamned candlelight dinner with Wayne." Immediately, I knew by the crew's vocal reaction that I had said the wrong thing. Within minutes, a "Candlelight Dinner with Wayne" signup roster was posted that included specific dates, crewmember names, and topics for discussion. Shortly after that, a special small table was set up with a white tablecloth and candles: Candlelight Dinner with Wayne was born. I very much enjoyed those dinners and the conversation that might not have occurred during my normal discussions with crewmembers. I came to the realization that the reason I did not eat with the crew on my second and third winter was not solely the desire to eat and read but that more and more I preferred to be alone. In thinking about this, I found myself desiring to be alone more and more and noted that even when home, I rarely ate with my wife.

While I did not eat in the galley often with my second and third crews, I always made it a point to stop by at most meals and say hello. During some mealtimes, I would try to interject humor or a public congratulations for a deserving crewmember. It was usually a light interaction, and I enjoyed being there with them. While I desired to eat alone, I did not want to totally estrange myself from my crew. I was able to interact with most of them during my walks through the station and their workspaces. I was always available if they needed to talk. They took me up on that many times, and I was always honored to be of help. Once, while standing and talking to someone in the galley at a mealtime, I overheard a conversation from a nearby table. Galley conversation could run the gamut of things from the cerebral to the absurd. On this occasion, the discussion concerned peas. It was lunch, and peas were being served. These were simply frozen peas that came in a plastic bag and had made an amazing trip of many miles to be on that table. The conversation was muted, with one crewmember simply stating that she really liked the frozen peas and others joining in in agreement, and it went on for some time. It struck me so hard later that this, my second-year crew—as superb a group as ever existed at the South Pole—could appreciate something as simple as a frozen pea from a plastic bag. I was humbled and in awe. One might take that little conversation as a minor occurrence, but I thought it so much more.

During my first winter, one of my crewmembers showed a somewhat obscure film called *The Chef of South Polar*. After viewing it, I bought a good copy of the film and showed it to my next two crews. I believe this film should be shown at the South Pole as a tradition. The film is a comedy of sorts. It is the story of a Japanese winterover crew at their Dome Fuji Station. There are several differences between their situation and the one at the South Pole, such as their station and crew are much smaller than Amundsen-Scott. There are also cultural differences. The film is in Japanese with horrible English subtitles that, at times, seem way off from what the character was probably really saying. Some things just don't translate well. Even with that, I consider the film the best representation of what one faces wintering at the South Pole Station. In the film, the main character is their cook. There is great emphasis on the crew's food, which the Japanese are masters of with both taste and presentation. A drunken birthday party is spot-on, and the terrible isolation and its effect on the crew is captured poignantly. There is a long-distance relationship breakup in which a young crewmember is left devastated and can do nothing but continue through the long winter. This is a very common occurrence during a South Pole winter. The scene that had the greatest impact on most of us was when the Japanese crew ran out of ramen noodles. With that terrible occurrence, life as they knew it changed for the worse. After having experienced a winter where there was a shortage of ramen noodles, there is a visceral quality to losing such an Antarctic staple. After the movie, several of us were hungry and headed to the galley for ramen noodles.

The South Pole Station has a hydroponic greenhouse that many of the crewmembers really enjoy. Prior to my first deployment, I had seen photos of it and had heard stories about it. I knew that it was bright, warm, humid, and that some of the crewmembers would spend time in there on a couch or chair reading, listening to music, socializing, and enjoying the atmosphere. As I had spent years in the tropics and was quite comfortable with heat and humidity, I thought that it would be a place I would be spending some time. In almost three years, I spent very little time there. I don't know exactly why, but at the South Pole the heat and humidity did not attract me. I would make daily checks of the system but was never drawn to stay. To me it seemed out of place to see the brightness and greenery and to feel the heat and humidity among the many miles of dead ice. I began to prefer the dead ice.

The greenhouse and what it produced was different every year and depended on the skill and desire of each crew. Some crews barely used it and others embraced it with a passion. It produced lettuce, tomatoes, cucumbers, and types of spices. It was not large enough to provide a major source of fresh food to the crew, but it was usually utilized for large formal dinners where everyone would have a salad of sorts. While I personally did not find it attractive, I was glad to see crewmembers who enjoyed spending time there. It was alive and so were they.

The term "dishpit" refers to the area in the galley where dishes and pots and pans are washed. It also refers to a duty that all winterover crewmembers perform. This consists of washing the dishes, pots, pans, and cleaning the galley to prepare for the next meal. Dishpit is done by the crew at all evening meals from Monday through Friday and for all three meals on Saturdays. On Sunday there is no formal meal service and volunteers will usually help clean up. From Monday through Friday, the steward does the cleanup for breakfast and lunch. He or she arrives at work early every morning to a mess and a mountain of dishes, pots, and pans. During the best winters, people will jump in and help the steward when they can. This is especially important toward the end of a winter when the steward and many in the crew are exhausted. To me, the steward was always the real unsung hero of any winter crew as their work is extremely repetitive and at times, with pots and pans having indescribable things baked to them, quite grueling.

During the summer prior to the start of my first winter, I was approached by several crewmembers who thought that galley personnel should be excluded from dishpit duties. There is some sense in that as the galley staff had worked all day in the kitchen and were then expected to stay and clean up after meal service. I gave this some thought, and it was easy to answer "no" to the request. The galley makes up almost 10 percent of a crew's total number, and as soon as they are excused, there could be cases made for others. The science people with their nonstandard schedules, the maintenance specialists with their long days, and others with their own special cases. I thought that the request to exclude the galley staff would lead down a slippery slope, and soon there would only be a handful of people on the dishpit schedule. I implemented the usual alphabetic dishpit schedule, based on crewmembers last names. It ran from A to Z, and all crewmembers, with no exceptions, were on the schedule. I explained the system at an All-Hands Meeting, and while it might have initially appeared inflexible, the opposite was true. I made clear that this was the schedule and that crewmembers were required to follow it, but they could trade dates. I encouraged the crew to volunteer to take the place of galley staff and others who may have had long days. This put the responsibility on the crew, and they handled it very well.

Dishpit was something that everyone said they would look forward to during their interviews, but later, few did. Most simply tolerated it and made the best of it, as it was a chance to do something a little different and they could play the music they liked while doing it. The only thing I liked about dishpit was finishing one shift and knowing that the next time I would have to do it was over a month away.

When crewmembers had face-to-face interviews there was a question designed to ascertain if they would be willing to perform the dishpit duties and other menial tasks. The question was something like "describe a time you had to do a menial task and how you felt about that?" Most people had good answers, especially if they had a military background. They would describe how they helped clean things, sweeping floors and doing dishes in a family setting. Every now and then, candidates had a different idea of what "menial" was. Some considered doing a certain type of report menial, and thus I would focus them on what lay ahead if selected. I made special notes when crewmembers spoke of how much they would enjoy it. They would usually state they thought they would enjoy the humidity and the change in schedule. Later, while at the South Pole, I found that most crewmembers did seem to be okay with it. They had been forewarned. It was fair as all crewmembers participated, thus there was not much to complain about. I do remember walking by the dishpit one night during the middle of winter and asking the fellow on dishpit duty how he was doing. His reply was immediate, quite curt, and quite obscenely phrased; he was not enjoying the experience. I remembered his specific interview and his bubbly response to the question. I immediately wanted to read back to him exactly what he had told the interview panel about how much he was going to enjoy dishpit duties. I did not do such as I thought that in his current mental state there was a real possibility of a pan or utensil being thrown my way.

On my dishpit days, I learned early on that the best way to do it was to have a team of people. Crewmembers helped me and in turn I helped them. I probably looked pathetic during my first dishpit and really appreciated their help. I thought it was going to be easy as I had dishwashing jobs in high school and enjoyed the hard and chaotic work with the servers screaming for dishes, glasses, and silverware. I soon saw that any skills I may have had years ago were now long gone. What I found most difficult about dishpit duties was the multitude of pots, pans, and the other kitchenware. There were many items that I could not identify purpose or storage location. The dishes themselves were easy, but the strange looking pieces of food service equipment encrusted with varying burnt-on coatings could be difficult to clean. My biggest challenge was that it took me a while to figure out where they were supposed to be stored after cleaning. I do admit to, during my first winter, putting some things back in any open area I could find if no galley staff members were around to show me where it was supposed to go. I think they are still finding some of those things.

On Sundays, there was no meal service. Hungry crewmembers foraged from a refrigerator that contained labeled containers of leftover food from the previous week's meals. There was also an option for staples such as ramen noodles, peanut butter, chips, cookies, or other things they had stored. I would set up the day before with sandwich meat, noodles, popcorn, and other food items that did not require actual cooking. I had no cooking skills and had spent many years at remote sites where cooked food was provided. I was always happy when Monday arrived and the regular meal service with real cooked food was once again available. Most crews had people who enjoyed cooking and would cook a few things on Sundays. This was usually well received by the crew. As there was no official meal service, there was no one scheduled for the dishpit on Sundays and the crews cleaned up after themselves. For the most part, they did a good job of not leaving a weekend mess for the steward to clean up when he or she started the duties on Monday mornings.

My weight at the South Pole was something I watched carefully, especially after the first year. During that first winter, I paid no attention to my weight and ate anything I wanted. I ordered through the summer mail other types of food, such as popcorn, which I coated with butter and ate in massive amounts. I really desired that, after being outside for any length of time. I started the first winter with approximately eighteen pounds of M&M's, which I ate within the first four months of winter. My thinking was that I was doing plenty of outside exercise and should burn many calories. I knew I was gaining some weight but did not mind the extra insulation and knew I would lose it when I left the ice. After winter, upon checking in at the Heritage Hotel in Christchurch, I was stunned at what I looked like in a full-length mirror with my clothes off. I had gained about twelve pounds. To most people this would have been only a mild cosmetic problem, but I liked to stay in shape and not at all soft. During the next two winters, in addition to my daily walks outside, I did a few miles in the gym on the treadmill and was slightly more careful with what I ate, especially during the last four months.

WINTER TRADITIONS

Yuri's Night is celebrated on April 12 and commemorates the great Russian hero Yuri Gagarin, who shocked the world on that day in 1961 when he became the first human launched into space. The South Pole celebration involves crewmembers dressing up in space-themed costumes they manage to cobble together from things in the arts-and-crafts room. Those who take the event more seriously will have brought costumes from home. In costume, they will congregate in a lounge showing space-themed movies and playing space-themed music, drink heavily, and ponder space. While it is all in good fun, some crewmembers are a little more serious about it as they are interested in joining the astronaut program. To date, several former South Pole winterover crewmembers have gone on to be a part of the NASA program, and several have been involved in the commercial sector's space program. There is a space station feeling at the South Pole in the winter. Once the sun has set and the windows have been covered, the station and the alien landscape outside feels like something out of a science fiction movie. Due to the extreme isolation, some South Pole winter crews have participated in NASA studies designed to understand the isolation that future Mars mission astronauts will someday experience. Yuri's Night is good fun and a reminder of the bond that the South Pole winter experience has with the space program.

The "300 Club" is a South Pole tradition that does not officially exist. It is not recognized or sanctioned in any way by the USAP or the NSF, but it is a reality. To become a 300 Club member, it must be colder than minus 100°F outside and specific tasks must be completed. The first is that the person will spend some time in the station's sauna that is set to 200°F. After a participant is sufficiently broiled in the sauna, wearing nothing but shoes or boots, they will leave the station, run around the Geographical South Pole marker, then run back into the station. Desiring membership in this exclusive club, the uninitiated winterovers anxiously await the first time the station experiences minus 100°F temperatures. On those days, I could feel a certain buzz and excitement around the station, as crewmembers mentally prepared for the ordeal. Personally, I found the event stupid. I have a fondness for harsh conditions but not for those that are contrived. To me, suffering can be noble and readily worth it, if it leads to real achievement, not simply a certificate or patch. It reminded me of the "double dip" done out on the Alaskan Aleutian island of Shemya. On Shemya Island, the participant runs into the Bering Sea, submerges in its frigid water, then makes a short run down a road to where it is said the Pacific Ocean starts. The participant will then run into that part of the ocean and submerge again. After completion, successful finishers are awarded a certificate that will forever commemorate their participation in the silly event. The 300 Club at the South Pole awards its members with certificates, stickers, and patches.

While I find the concept of the 300 Club stupid, it is a South Pole tradition and I do think it serves to further bond crewmembers. In three winters, I did not do anything to stop it, but I made sure our doctors were ready and participants were as careful as possible during the event. As grueling as the event sounds, participants have told me that they complete it with such speed that they do not experience the cold they thought they would. Historically there have been few injuries, although I did hear a story of a prior year where a doctor who was a participant became disoriented in the outside darkness and spent too long outside the station. This

resulted in frostbite to his body's frontal region, and with the scab-
bing, he could not bend or sit down for a while. All aspiring 300
Club participants were always informally notified that there would
be no reported injuries.

On the days when the 300 Club miscreants were preparing for
their big adventure, I also prepared for something I enjoyed. One
of the reasons the 300 Club does not freeze its participants to death
is that at the South Pole when the weather is colder than minus
100°F, it will usually be relatively calm. Thus, there will be little
windchill and visibility will be good. On those days, I liked to get
out and take a longer-than-usual walk on my grid-south route. I
remember well my first walk in minus 100°F temperatures as I could
hear the screams of the 300 Club participants in the darkness on
their way to Antarctic glory. Walking out of the station, I followed
what had in the summer been the SPOT road. Winter storms had
months before obliterated the road, but my own personal trail,
which I started during the summer, still existed in places. It was an
eerie feeling that first year, out a mile or more from the station in
such a lethal environment. It was extremely dark, and the airfield
distance markers were frozen over and hard to find. I knew I was
close and had to do grid-like walking patterns to finally be able to
find them. After several winters I was much more comfortable, but
I was never complacent and always realized the potentially lethal
situation for what it was.

Every year the winter crew designs a new Geographical South
Pole marker. This marker is placed on a metal pole, planted in the
ice, and remains in place at the Geographical South Pole for the
coming year. The design of the marker is open to all winterover personnel in the form of a contest. There are a few basic rules to include
certain things in the design that will make it an actual geographical
marker. It must include exact wording that notes geographic location and have National Science Foundation engraved on it. Once
those requirements have been satisfied, the overall design is open
to the designer's creativity and the ability of the winter South Pole

machinist to make it. Once all designs are completed, the entire crew votes for the winner. The winning design is then submitted to the Denver office and the NSF for approval, which is needed to keep things from getting outrageous or distasteful. One year, after explaining the design process to the crew at an All-Hands Meeting, I told them that designs had to be in good taste and couldn't include something like a penis. I was not surprised when soon afterward a large metallic mockup of a penis pole marker was presented to me. I kept it in a box in my office, and at some point prior to station opening, I told our machinist to destroy it, or to at least change the crew year engraved at its base from ours to that of the of the crew that would be relieving us.

There is a glass case on the first floor of the station that displays the previous pole markers. These go back many years. The first designs were quite crude, with several that are simple pieces of copper or piping. As the years progress, the markers become more sophisticated, and at some point, they turn into works of art. As works of art, they do begin to lose what they were originally designed to be, simple geographical markers that indicate 90 degrees south.

During my three winters, I tried to tactfully and without any type of real pressure, steer the crews away from sophisticated, over-the-top designs and advocated for something simpler. During my second winter, I nearly succeeded. Several submitted designs were quite retro and similar to the simple copper-pipe concept from years prior. That direction was quickly upended by a change that the USAP made to the organization's logo, which includes a map of Antarctica. The change to the logo removed a small star-like emblem from over the South Pole. As South Pole winterovers always understand that they are the center of the universe and certainly the center of the Antarctic continent, action was needed to rectify the errant change. The design that easily won that winter's pole-marker contest was the three-dimensional embodiment of the star that had been removed from the South Pole USAP logo. It was a beautiful design and a symbol of the creativity and mindset of a South Pole winter crew.

On my third winter, we learned of the off-ice death of a machinist that had wintered several times and was known for his beautiful pole markers, both in design and execution. He was supposed to be on our winter crew but at the last minute, had a PQ problem that prevented it. Learning of his death hit the crewmembers that knew him extremely hard. We decided to hold a remembrance for him in our galley early one Saturday evening. We first served what he enjoyed most, prime rib. That night, I ate with the crew. It was an incredibly solemn event, which I tried to slightly lighten up by telling a funny story he starred in. I knew that given the choice between solemn or humorous, he would have wanted the latter. Several people spoke after I did, and we had a toast to him and his eternal memory. I broke my own rule and had a drink with the crew to remember him. He had written me a message in February expressing his sadness for not being on the crew but said that he was working on clearing the PQ issue up and hoped to be on the next. He died and never was to return to the South Pole, but he will live on for eternity as his beautiful pole markers will someday end up in a major museum, if not stored at the South Pole forever. I used to run into him on occasion outside on the flag line that ran to his shop, which was in one of the science facilities. He had a very distinctive voice and accent, and even in darkness and the worst weather, where I could not see well, I was able to recognize him. After his death, when I was out in the darkness and walking that flag line, I called to him several times to see if he was there. I know he loved the South Pole, and I thought he might have wanted to spend eternity with us. I did not find him and hope he rests eternally in a sunnier and brighter place.

Another South Pole tradition that occurs in the winter is the taking of a crew photograph that will hang in the long hallway on the second deck. The photos started with the first winterover crew in 1957 and go up to present day. As you walk down the hallway, starting with the 1957 photo, you see small, framed black-and-white photos, the first few being taken indoors. As you proceed, the photos get larger and more colorful. There are photos that contain signs of how that winter went. If you notice many "not pictured" crewmembers, there were problems. Having many missing crewmembers in a photograph is an indication that all was not well with that winter, and crewmembers were making a statement by not wanting to be photographed with the crew. There are people that simply do not like being in photos so a person missing here and there is a normal occurrence. One photo that really stands out is contained in a large metal, circular frame. The frame has bolts running through it that correspond to the number of crewmembers that year. Four of those bolts are black and signify a horrendous year where mayhem ensued. Hiring over the phone, which was the norm at that time, resulted in several crewmembers being selected that were not fit for such an assignment. At the beginning of that winter, a small group of miscreants led by a vial blowhard, who was a last-minute hire, engaged in a series of minor vandalisms and other unpleasant actions. Weak leadership then gave the group of four the forum they needed to cause disarray and much unrest among the crew. These miscreants are for eternity known as the "black bolts."

One of the many acts of their minor mayhem involved one of the miscreants breaking a galley plate over the head of a crewmember who dared to complain of her idiocy. The broken plate was contained for years afterward in a Ziploc bag in the area manager's office. During my second winter, at an All-Hands Meeting, I used it for a winter morale boost. I brought the bag of plate shards and asked for volunteers to try to reassemble it. I had noted that our crew had several people quite adept with puzzles. It was then reassembled by a very skilled crewmember in a manner somewhat like what one sees with broken and restored ancient Egyptian or Greek pottery that you find in museums. It was once again a plate, but with small gaps from missing bits. A wooden stand to hold the plate was crafted and the specific crew's year was put on it. I placed the plate and its stand in one of the stations glass display cases. During station opening, I proudly showed it to the incoming summer management staff as one of our winter station improvements. I also gave credit to my crew for helping to heal the past. The plate was removed from the display case shortly afterward. I assume because they felt it had a negative presence. I do agree that it would have led to embarrassing questions should a summer visitor want to know more about it. Many lessons learned from that winter resulted in changes to the hiring practices, most notably the requirement for face-to-face interviews. These changes helped alleviate the possibility of ever hiring such a group of miserable creatures again.

I would often walk that long hallway, starting with the first years and thinking of all the people who had previously wintered. I would think about the vast number of stories that would be generated by all those people with their many experiences. There would be quite the range, from great triumphs to great heartbreak and many things in-between. I would consider how many of those people had died, as I had gotten several reports of deaths that had occurred for the men on the first winterover crews. I used those photos to try to motivate my crews with the fact that for as long as the United States had a presence at the South Pole, they would be there in a photograph. It then came down to how they wanted people to remember them. Did they want to be a black bolt? I never met anyone who did, and I never had a black bolt on any of my crews.

Near one end of the long station hallway, there is another group of photos. No one wanted to be included with that exclusive group. The photos honored men who had died at the South Pole. There were four photos, and they all had a short synopsis of the person and the way they had died. One had died of natural causes and three had accidents of sorts. The photos served as a powerful reminder for the need to exercise great caution when working at the South Pole. At any of the formal dinners, a special table is set for the four, to honor them. They are with us always. During my third year, I looked at that wall differently and thought, if I did die, especially of natural causes, it would be an honor to have my image reside there for eternity. There was a space at its far end that I thankfully did not end up occupying.

The photos of the dead somewhat drew me, as I knew something of death myself. Many years prior, I had been listed as killed in an airplane crash in the Amazon jungle and had a framed newspaper at home documenting the event. I had missed being killed in that crash, which took over seventy lives, due to my Spanish language skills being poor and not understanding when to be at the airport. I was on the flight manifest and was supposed to be on it, but not understanding what I had been told, I had hiked across the frontier into Brazil and made camp. When I returned, the flight had departed and all were dead. Rather than diminishing in my mind through the years, the event became more powerful as I began to consider all possibilities. I had read and was haunted by Ambrose Bierce's short story "An Occurrence at Owl Creek Bridge." Out on the ice, I pondered this, and whatever the truth, I felt quite alive.

For some reason, during my first winter, my moustache became a focal point for the crew. I had a moustache since leaving the Marines many years prior. Through the years, I only removed it once. That was when I served as a Defense Department contractor in Iraq. I did it then as I was wearing a U.S. military issue uniform with sidearm. For me, the mustache was a permanent fixture and even a sort of trademark. Early in the winter, crewmembers began trying to grow their own mustaches, with some amusing results, as the younger members tried and grew things something like a moustache but without much success. Once, a young crewmember had shown me his new growth and told me he was now beginning to look like me. Instinctively, I responded with a comment I regretted immediately. I told him that his moustache was not the same as mine unless he had spent time with cannibals, parachuted out of a jet, and had smuggled guns into third world countries (for personal protection, but I didn't tell him that). I regretted the comment immediately as it exposed a character flaw concerning my own prideful ego, which I can never quell but can at least try to conceal. I had experienced copycat types before, with aspects of my dress and accoutrements, and was always uncomfortable with it. This was just good fun, and I should have been more gracious with the comment. The moustache symbol then began to appear on other things, such as a large banner used for social events and a flyer based on a popular TV series. The crescendo was the symbol of the moustache emblazoned in a carpet tile on the galley floor. I held a minor ceremony for that at an All-Hands Meeting. I knelt and had photos taken with it. I laughed at the thought that it was similar as to what happens in Hollywood with the actors and their stars on the Walk of Fame. While I found the attention quite amusing, I also saw it as an honor.

Some South Pole winters develop themes official or unofficial. Sometimes it is a shared hatred or disrespect for the WSM. The WSM is a difficult position for most, especially if they have not been in leadership roles. It can also be difficult for those that have military backgrounds and think they can lead a crew utilizing what they learned from the military. South Pole winter crewmembers are, in general, eclectic individuals who lean left politically. They do not respond well to military leadership. The crews may have a few military veterans, but most of those people have had their fill of military leadership and want none of it now. I remember a specific example of that with a WSM who had retired from the military. In the heat of a moment, after learning that something he had asked to be done had not been done, he shouted that he was giving the individual a "direct order." The receiver of the direct order was also a veteran but did not comply and found it amusing. This enraged the WSM, who quickly departed the scene, slamming a door and leaving a hole in the office wall. In this case, the WSM was a true leader but was with the wrong people to use that style of leadership. There have also been past WSMs who had little leadership experience but had previously wintered at the South Pole in other positions. Denver South Pole management sometimes selected WSMs that had past winter experience in other positions. The idea was that with a person's past winter and technical experience, they should be successful. These people usually were not leaders, and for those years, the Denver office ran the station from afar. One challenge with this type of promotion is that the person may still be seen by the crew as what they were in their former position. In addition, the person may revert to what they know best from their previous position that did not include leading a South Pole winter crew. While prior South Pole experience would seem a good thing, sometimes coming from outside the program and being a bit of an unknown has its advantages.

During my three winters, I wanted to have crew mottos or symbols. I wanted these things to set us apart and help develop a sense of pride among the crew. During my first winter, the symbol was the moustache, which the crew had a lot of fun with. The motto was an inscription on our geographic pole marker that came from the great Antarctic explorer Ernest Shackleton: "By endurance we conquer." I think this is a fitting motto for any South Pole winter crew. The pole marker that the phrase was inscribed on was a thing of beauty, but I almost disfigured it by a well-intentioned act. That winter, I had brought to the South Pole a small grey pebble, which had been given to me, that came from Ernest Shackleton's grave on South Georgia Island. One day, while cleaning my room, I found it on the floor and assumed it had fallen out of its container. It seemed slightly smaller than I remembered it, but I had not seen it in some time. I talked to the machinist who was making the pole marker: as we would have Shackleton's motto on the marker, why not drill a small hole in it and insert the pebble from his grave? He said it was possible, but thankfully, I did not follow up and have it done. Some months later, I went to the small jewelry container that held the Shackleton grave pebble and other prized objects. I found the stone there, and it was larger than the one I had found on the floor, as I had thought. The pebble I had picked up off the floor in my room was simply a pebble that had come off my shoe. It probably came from a Denver hiking trail. Thankfully, that meaningless pebble was not entombed in that year's pole marker!

I was extremely concerned about my second winter and worked hard to put together a crew that included many veterans. I was once again part of the rigorous winter crew selection process that occurred in Denver, but now, I had experienced a South Pole winter. I could now quite readily and with some detail answer what minus 100°F felt like, and I had spent over 1,500 miles outside. I was able to attract several veteran winterover crewmembers from my first winter back to the program. I also looked for other veterans from other winters. This effort resulted in a crew where nearly 50 percent were veterans of past South Pole winters. Around 70 percent had previous USAP experience. All but two of the science personnel had previously wintered. In addition to having a large number of veterans on the crew, each crew member was extremely qualified. I felt that crew had the potential of being the most qualified crew in South Pole winter history.

I wanted to give them a motto that had a deep meaning and an edge. "We ate the dogs" was something that hit me while out walking. At our first meeting, just prior to the station going into winter mode, I revealed it. I explained how in 1911 Roald Amundsen and his gallant men had reached the South Pole. That achievement had been a combination of meticulous planning and their acceptance that they would need to do something that many would find unpleasant in order to reach that goal and safely return. They would eat their dogs. I explained to the assembled crew my goals for our winter. First, that we would not have a serious accident or injury and that they would return home whole and unharmed. Second, that we would return the station to the summer people in better condition than we received it. Third, that we would have no serious personnel issues during our winter. Last, that we would come out of this as a strong crew and proud of what we had accomplished. I explained that we should be willing to do anything it would take in order to accomplish these things, and that we would "eat the dogs," metaphorically speaking. I did not sense from the assembled group an immediate love for the phrase. In retrospect, it probably sounded harsh, especially for anyone who loved dogs, but I wanted to get a point across, and I wanted it to have a harsh edge. On the tubular base of the Geographical South Pole marker that crew made, the crewmembers' names are listed and that phrase is inscribed for eternity. That was a phenomenal winter crew, and whether or not the phrase was embraced by all, they truly ate the dogs.

I did not personally select most of my last crew as I did a back-to-back winter and the hiring took place at the Denver office while I was at the South Pole. I did not know the crew and was somewhat concerned as they were a younger and much less experienced crew than the crew I had just completed the year with. Wanting to catch their immediate attention upon their arrival, I had introductions with them that, I learned later, some thought were harsh. I explained to them that it was summer and not winter yet. I told them that some of them might not be around for winter. My reason for doing this was that I would be leaving for much of the summer to get the mandatory USAP forty-two days off-ice between winters and complete my required PQ. I knew the summer was a dangerous time for winter crewmembers as they could get caught up in summer antics and then be terminated. I would not be there to help them navigate through this, so I wanted to leave them with a heavy caution. Even with the caution, after I departed, several crewmembers were lost to the summer and its antics. Upon my return to the station at the end of summer, I had an All-Hands Meeting for my winter crew and decided to improvise. After one look at the assembled, complete crew, I decided not to be as dramatic and forceful as I had planned. Before me was not the veteran group of "dog eaters" from the previous year, but a much younger and untested group, so I held off, at least for a while.

Several months later, during my speech at the Sunset Dinner, I felt I knew the crew well enough to reveal a little more about expectations and show a little more of my true self. I quoted one of my favorite passages from *Moby Dick*. In this passage Captain Ahab explains he will not be "swerved" from his quest to hunt and kill the white whale: "The path to my fixed purpose is laid with iron rails whereon my soul is grooved to run, over unsounded gorges through the rifled hearts of mountains, under torrent's beds unerringly I rush. Naughts an obstacle, naughts an angle to the iron way." I offered that gem to the crew, specifically "the iron way," as a possible crew motto. That phrase is burnt into my being, and I have lived it. I can literally smell and taste that iron as I hurtle down my life's path to its eventual oblivion. On that evening, looking at that crew, "the iron way" seemed a little off. Every crew is different and with proper leadership, can accomplish many great things. That crew faced unprecedented off-ice hardships, and while they did not deal with challenges with Captain Ahab's fanaticism, they were successful. Out on the ice in the darkness, their own Captain Ahab roamed.

I loved the Ahab character, and although toward the end of the novel, I saw him reduced to a crazy old man, I found him beautiful. To be more precise, I found his potential beautiful. He was an accomplished veteran sea captain who bore scars of a horrific encounter and had an unfortunate given name. He was isolated from his crew, but his presence was felt. He was a brilliant orator who rallied the crew to participate in his obsessive quest to find and kill the white whale. He had a beauty that was similar to another epic figure who had fallen from grace eons prior. That beautiful figure carried the light but chose the darkness. The book is a cautionary tale of obsession and the price paid for having priorities that did not place people and their lives as the most important thing in the universe. As a sea captain, Ahab's primary duty was to his crew and ship. Their health and safety should have been in his every thought. In his case, he let his obsession kill them and destroy his beautiful ship. My obsession was not killing a white whale, it was the health and safety of my crew and the beautiful station. That was my sole, fixed purpose.

Another South Pole tradition is the replacement of the twelve national flags that have flown for a year at the ceremonial South Pole and the American flags that have flown on the station and at the Geographical South Pole. The twelve national flags at the ceremonial pole are the original signatories of the Antarctic Treaty. The replacement is done when the sun first comes up in September. By that time, the flags are all quite tattered from the cold and incessant Antarctic winds they have endured for a year. A group that has an interest in the event gets together and walks out to the ceremonial and geographical poles. The old flags are taken down and new ones are hung. The old, tattered flags are highly coveted. They are brought into the station and, for most crews, at the Sunrise Dinner, they are raffled off. As the wsm, I had the authority to distribute them any way I saw fit, and for me, using a raffle system where everyone had a fair chance to win a flag was the best way. I did make two unilateral decisions concerning the flags that were supported by tradition. I awarded the National Science Foundation, nsf, flag to the science person that had served as the station science lead, ssl. Unfortunately, that flag, which was slightly smaller than the rest, had become so tattered that nsf said only ns, as it had been reduced to nearly half its original size as fragments had been blown to the winds. As the wsm, I had my pick of any flag I wanted. Most wsms knew they were only going to do that single year and took the U.S. flag that had flown at the Geographic South Pole. This is the most coveted of any flag. There are two other U.S. flags that are outside. One flies at the ceremonial South Pole and the other on the station building itself. During my first winter, and because I knew I was coming back for another year, I did not pick the U.S. flag but rather, the Norwegian. The tattered Norwegian flag was then signed by all crewmembers and had our crew patch sewn on. During most winters, the crews will have patches made that are specific to them. The process is very similar to what nasa does with each individual space mission. That flag with crew signatures and patch made quite the trip.

At the end of my first winter, I arrived home to Texas to a mess from Hurricane Harvey, but I still had one thing to do to finish my South Pole winter. My wife, Melissa, and I journeyed to Oslo Norway to the Fram Museum, which houses Nansen and Amundsen's famous ship *Fram*. *Fram* means "forward" in Norwegian. The ship is famous for its drift north, near the North Pole, with Nansen and later, for Amundsen's South Pole journey. It is a special ship designed for the ice and has a unique, very rounded and strengthened hull. This design allowed the ship rise out of pack ice once it had been iced in rather than be crushed by it, as the USS *Jeannette*, Shackleton's *Endurance*, and many other polar ships had been. Now, for eternity, the great ship *Fram* is housed in a covered museum in Oslo, Norway. The plan was to visit it with my tattered, signed, and crew-patched Norwegian flag, tour the ship, and get some photos with the flag. In essence, something important from the South Pole and all my crewmembers would be with me.

My wife and I arrived in Oslo, Norway, on Thanksgiving and walked through snow and ice from our hotel to the museum. It was an odd walk, being again in a cold and frozen place but not with the type of clothing I had worn for a year. Near the museum, we came across the full-scale, bronze replicas of Amundsen and his men standing outside wearing their polar gear. It was overcast and snowing lightly, and I touched each man's face. Prior to our arrival, I had sent a message about our visit; a museum representative then gave us a superb tour of the ship and museum. I tried to listen to the tour but was overpowered by my thoughts and imaginings of what this ship had experienced and of the men who had sailed on it. I imagined them walking the decks, eating, talking, sleeping below, and going on to make polar history. Several photos were taken, and later I heard a comment that in the photos I had not smiled. I could not smile. To me there was nothing remotely amusing or even happy about the visit or ship tour. I was not at all jovial. The experience was to me a great honor, quite serious, and simply overwhelming. I gave it much thought afterward.

At the end of my second winter, I chose the British flag at the flag ceremony, as I knew I would be returning for my third and a back-to-back winter. I planned the third winter to be my last, and I would at that time choose an American flag. The crew signed the tattered British flag, and our crew patch was sewn on. My wife and I then journeyed at Christmastime to Dundee, Scotland, where Captain Scott's famous Antarctic ship *Discovery* is moored and there is a museum. I had sent a message explaining that we would be arriving, where I had been for the last year, and what I had with me. The museum people were incredibly gracious and provided a wonderful tour of the museum and ship. Again, as on the *Fram*, I could not fully concentrate on the tour as my imagination was running wild as to the men who had been on that ship and where it had been. I was overwhelmed with being there and the great honor of representing my South Pole winter crew. I only wished they could have been there with me.

For my third and last winter, I chose the U.S. South Pole Station flag. Long before the flag ceremony, I had debated which flag I would choose, the flag that had flown on the station itself or the more coveted U.S. flag that had flown at the Geographical South Pole. I would look at the station flag after returning from walks and had seen it tattering over time with the intense cold and strong South Pole winds. I thought it more appropriate to choose that flag, as it was my station and my ship that moved with the ice. Still, I considered choosing the U.S. flag that stood at the Geographical South Pole near the place Roald Amundsen and Captain Scott had stood many years prior. In the end, the decision was made for me because at first sunlight, when we went to change the flags out, it was discovered that the flag at the Geographical South Pole was gone. It had blown away with its wooden pole during a winter storm. Efforts were made to locate it by excavating near where it had last stood. We felt that with the weight of the flagpole and its lack of aerodynamics it could not have blown far. The efforts to locate the flag were to no avail, and we made plans to have incoming station personnel continue excavation during the summer after we had departed. Thus, the decision of what flag I should choose that year was made by the severe Antarctic winds. I feel that harsh environment helped provide me the right choice. That flag will be making a special trip too, and once again, my crew will be with me.

OUTSIDE

In the winter, I had three reasons for being outside every day. The first was that I needed to be able to check outside conditions for my crews as they would be out in it. The science personnel had a long walk to their telescopes, which occurred regularly and during an emergency. The maintenance personnel had their daily outside rounds, which could be quite rigorous. The logistics personnel could be digging needed equipment or supplies out on one of the storage berms. The communications personnel would be, on occasion, out at their satellite domes, which were over half a mile away. Winter and all but the worst conditions did not stop this from happening. As the WSM, I was authorized to restrict outside travel and in the case of truly awful and dangerous conditions, prohibit it. It seemed to me that a real leader should regularly get outside the nice warm station and understand the conditions the crew was facing. I never led from a desk. In three winters utilizing sensible outside travel procedures, I never had to prohibit travel and we never had anyone lost. The second reason was exercise. While I did utilize the station's gym, especially in my second and third winters, being outside and moving was vastly preferred. The third reason happened after I had spent considerable time outside. I enjoyed being out there. I enjoyed being alone and away from the station. I enjoyed having some level of challenge and harshness. I realized it would never be anything comparable to what my great heroes had faced, but I also knew that none of them had ever experienced the conditions of a winter at the South Pole.

I enjoy true harshness. Prior to my arrival at the South Pole, I had several experiences that would support that fact. One was a trek from the interior of Indonesian New Guinea, from the highland capital of Wamena to the Asmat coast. This occurred many years before I arrived at the South Pole, but there were lessons learned that I never forgot. Taking two highlanders with me, I crossed the mountains that surrounded the highland plateau. In addition to my two highlanders were other villagers who carried loads of supplies for certain distances, until they would refuse to go farther, and new villagers would take their place for the next section. This was necessary because tribal wars between villages would make further travel too dangerous for them. Over the mountains and down the wet, rainy side we went, until reaching the swampy areas that ran for hundreds of miles. From there it was dugout canoes, and finally, boats that got us to the coast. It was at times gruesome, muddy going, with leaches; snakes; crocodiles; and finally, disease. For navigation I used a compass and an old aviation map that still had blank spaces. I never knew exactly where we were. What I knew was that we needed to go south and we would eventually hit a river that would get us to the coast. My order through the dense jungle was simply, "*Selatan! Selatan!*" (Indonesian for south.) At the coast we were stuck for several weeks awaiting a flight back to the highlands. During this time, both the highlanders contracted malaria. They were placed in filthy beds, and I nursed them back to health with the little medicine I had. Malaria is an insidious disease, and New Guinea has some very dangerous varieties. It never really goes away, and there can be episodes of recurrence later. Although the symptoms dissipated, they took that disease back to the highlands with them. I caused this with my vanity, obsession, and will to accomplish such a trek. I have never forgotten it.

At the South Pole, I had two routes that I walked. The first was the one-kilometer flag line that ran out to the South Pole Telescope,

SPT. The flag line had small red flags on bamboo poles planted in the ice every twenty feet or so. There was also a small reflective strip on the flags that would catch the light of a headlamp. During the worst storms with blowing ice crystals, one could navigate from flag to flag and not get lost. The scientists assigned to the ICL, SPT, and MAPO projects and our maintenance personnel needed to be able to get to these facilities in most any kind of weather. The flag line provided a safe route in all but the absolute worst weather. On many days, I would do multiple laps out on it and would average three to four miles. For me this was nothing very adventurous, simply exercise and time outside. What I enjoyed most about this walk was my ability to let my mind concentrate on other things, with no real worry about navigating and getting back. Even with a flag line, there were winter storms that turned everything into a mess, and visibility would be quite low. I well remember a walk on that flag line in terrible visibility, during which I ran right into one of my crewmembers who was out on maintenance rounds. He asked me where the station was. This was slightly unnerving, not for my sake but because one of my crewmembers had become so disoriented. It struck me hard how much worse that could have been if he had lost the flag line and begun wandering. This was during a condition where the ice crystals were obscuring ground visibility but you could see the stars above. I showed him a bright star and told him the station was under that.

My second and preferred route was grid south. At the South Pole every direction is north, so a grid system is utilized for local navigation. Grid south was roughly where the SPOT Road had been in summer but was now obliterated by winter storms. From my room, I would enter a vestibule and walk out the large freezer-like door at the back of the B wing. Headlamp on, unless the moon was very full, I would walk down a couple of flights of exterior stairs that led to the icy surface. At the start of a walk, there was a minor chill as cold permeated the gaps in clothing and face coverings. This would diminish with forward motion. I would scan grid south, seeking the red lights on the satellite domes as my initial aim, but trying to stay to the right of them. The farthest dome was around .7 miles from the station, and after I reached a point to its right and was even with it, I would need to continue on and walk at a slight right angle to seek the one-mile skiway marker. In total darkness this could be difficult to find as its black, painted surface was now covered in frost. If it was a clear night, I would pick out stars that I thought were in the right direction and head toward them. Near that one-mile skiway distance marker there was an unlighted meteorological station. If I could find that, I could use it for orientation. After finding the one-mile skiway distance marker and looking back at any red station lights I could see, I would orient myself and choose a star that I felt was roughly above the mile-and-a-half skiway distance marker, which would be slightly left. I would walk in that direction, and sometimes there were remnants of footprints from my earlier walks. It could be quite dark if there was cloud cover that obscured the moon and stars. I would choose my steps very carefully over the uneven sastrugi. My pace was slow, methodical, and I did not want to fall. I knew that falling could lead to an injury that could become a serious issue. From the mile-and-a-half marker I occasionally, in the darkness, journeyed out to the two-mile marker. Finding that could be quite a challenge in total darkness. During my first winter and to a lesser extent my second winter, my senses were on high alert on the longer walks, and I will admit to a sense of nervousness. In reality I wasn't that far from the station, but it felt like I was a world away. During my last winter, I felt nothing like that. I felt one with the ice, but I was still extremely careful.

I had learned several important lessons on my walks during my first summer. Those walks were relatively easy, with twenty-four-hour light and cold that was manageable. I started out very carefully with winter walking as I had never faced such a combination of cold and darkness. One of my concerns, which had arisen during my summer walks, was that I was the station's incident commander, IC. In the event of an emergency at the station, I needed to move rapidly and take my place with the emergency radios in the communications center to coordinate the response. During a summer walk on a Sunday, which was our day off, I once tried to coordinate an emergency response by radio when CO_2 was inadvertently released in our emergency power plant during alarm testing. I was three miles from the station and my radio went in and out during the action. I ended up with slightly frozen hands as I could not use the radio while wearing thick gloves. I was furious upon my return to the station as the discharge should not have happened, but I did realize something needed to be done to ensure that the role of incident commander was immediately filled when I was away from the station. I developed an informal procedure with my ERT leaders that should I be too far away from the station to get back in time, someone I had previously delegated to fill in for that role would respond. It was important that people knew my location, so I developed a sign-out board to specify where I was. The board changed a little each winter as I added new details, such as my outside mileage to date and, more specifically, where I would be. Fortunately, this was never an issue during any of the winters.

By my second winter and after having walked over two thousand miles, it became a very natural event. Dressing for walking outside was a ritual. I always thought that preparing myself for the walks was akin to preparing for scuba diving. It started with two layers of long underwear, the inner a synthetic nylon type and the outer a fleece. I would usually choose the military Generation III Extreme Cold Weather parka and pants. On the coldest and clearest days, I would wear the wolfskins. Covering my head and most of my face would be two balaclavas. I wore a bandage over the bridge of my nose as I had found that area susceptible to frostbite if there was any wind. Over the balaclavas, I wore a knit hat that was a mixture of wool and possum fur. Around my hat was my headlamp, with a red light and an electrical cord to the battery pack worn on my waist. I would pull the hood up and over my head with the head-lamp protruding from the front. At that point, there would be just a slit large enough for my exposed eyes. My gloves were a silk inner liner covered by green wool military mittens, which had a trigger finger that, fortunately, I did not need. Over these, on the coldest days, I wore the large "bear paw," USAP-issued mittens.

For boots, I preferred to use my moosehide and canvas mukluks. These were extremely lightweight and comfortable. On the coldest days, I wore blue Asics FDX boots. My boots made a distinct crunching sound as I walked on the ice crystals. After walking many miles, I noted that each type of boot made a different sound while treading on the ice. The mukluks were incredibly quiet. They were my preferred boot for that reason and for their light weight. One warm night, with temperatures in the minus fifties Fahrenheit, I decided to wear my specialized summer running shoes with the ice treads. I thought they would be warm enough and looked forward to their light weight and traction. Immediately after I started walking in them, I noted a loud crunching noise as the ice tread pierced the frozen sastrugi surface. I had never paid any attention to the noise when using them to run during the summer, but now the noise seemed intolerable. I returned to the station and changed them out for the mukluks. I never wore them again.

I carried two compasses, two headlamps, and a charged radio. I would check my face prior to leaving the station to see if I had things lined up correctly as once outside, it was difficult to adjust anything. On one occasion, I was just outside the station and felt an immediate cold at the bottom of my legs. I had not pulled the military ECW pants down over my boots. On another occasion, I had forgotten the bear paw gloves and within seconds recognized that. Another time I forgot the wool and possum hat. I would check the weather prior to exiting the station, and most things I wore stayed the same every day. The only exception was the need for a wool Shackleton sweater when temperatures were in the minus 90°F and colder range.

I would form my mouth into an "O," like blowing into a straw; exhale heavily; and focus my breath forward into the balaclava, where it would cover my mouth and nose at the start of a walk. My goal was to get my balaclava face covering to freeze up with the water vapors I was breathing out. I found that after the mask froze into a cake of ice, it kept my face warmer. Out in the frigid stillness, my breath and the crunching of the ice as I walked was all I could hear. Being dressed as I was, I was quite safe and insulated from the cold. Out on that ice cap some distance from the station, I was truly alone and no longer part of the world. This was what I was truly seeking. I did it so many times and for so long that it became normal.

My favorite time to walk was when it was extremely cold, in the minus nineties Fahrenheit and colder, and I was heading grid south. Usually, that meant little wind. I was extremely careful and aware of the fact that if I had any kind of serious medical emergency I would die there. I had no reason to believe something as catastrophic as a heart attack awaited me, but at my age, it was at least a possibility. I thought if something so serious did happen, I would rather die out on the ice than survive in a reduced capacity and burden the station with an attempt at a winter air evacuation. I wanted nothing to do with being the subject of such an event that would put other lives at risk. In the event that such a fatal medical emergency did occur, there was a note on my sign-out board that said, "If no return look for frozen lump when the sun returns." It looked humorous written that way, but I was quite serious. My HEO knew my basic route south, so finding me after such an event would not be difficult. No one would be put at risk during the retrieval. What was more likely to happen was a simple fall over the uneven sastrugi, with a minor injury resulting, so I chose my steps quite carefully. My progress on this route was usually quite slow, about two miles an hour. I fell a few times during the winters and afterward would always look at the spot closely, attempt to determine what had caused the fall, and try not repeat it. During my few falls, I never had any kind of hard landings and would be mildly amused. To me, to have to call on the radio for rescue would have been nearly as bad as just dropping dead and becoming a block of ice. I did not want my crew to be burdened by a search or any kind of a rescue and was always extremely careful with each step. After so many miles outside in the dark, I became quite adept at what I was doing but was never overconfident.

One of the questions I often received while home between winters was "how did you handle the bathroom situation while out walking, especially at some distance from the station?" I think people thought that due to the extreme cold, it was impossible to do anything to relieve oneself. I can say, for me, it was never an issue. There was a very slight learning curve, but the routine for urination was simply to have the wind at my back, gloves off down to liners, and go! It was a quick process and was never any kind of problem. I am sure I contributed a few gallons of myself to the ice over the three winters. The other procedure was slightly more gruesome, but I became quite adept at that, too.

One memorable walk occurred with the temperature being minus 104 Fahrenheit. It was a very dark night with no moon and little wind. I walked my grid-south route and could not find the one-mile skiway distance marker. The painted, black face of it was frozen with a white, icy covering and it did not stand out in the darkness. Unable to find it, I continued grid south, using stars and following what I thought to be the right route to the mile-and-a-half marker. When I arrived at where I thought it should be, I could not find that marker either. After spending some time looking, I decided I had reached my goal in distance but was slightly frustrated that I could not find the marker. On the way back to the station, I ran into the meteorological station, which was a mile out. Realizing that the airfield one-mile distance marker was near that, I quickly located it. Now somewhat oriented and troubled that I had failed to find the one-and-a-half-mile distance marker, I headed back out to try again to locate it.

Knowing now exactly where I was, I used stars to guide me to where I thought I would find the marker. It was very dark, and

again, when I reached what I thought was the correct distance, the
marker still alluded me. I devised a new plan and started walking
in a grid pattern in the hope of running into it. I was walking in a
box pattern, orienting on a star that I felt was in the proper direction. After several hundred yards of walking and still no luck with
the location, I had spent much longer outside than I had originally
intended. Then a minor problem occurred. I had to shit! There was
no possibility of making it back to the station, so I did what unfortunately had to be done. When people press for details, the only
thing I will ever say to describe the act is the word "quickly." With
that bodily act, I attained what will probably be the closest thing
to immortality that I will ever achieve as that small, buried part of
me spends the next couple of hundred thousand years on the moving ice on its way to the Southern Ocean.

That type of thing happened a few times when I was some distance from the station, and it was not something I preferred doing,
both for the discomfort in the process and because I did not really
want to leave such a deposit unless absolutely necessary. I had
noted what summer tourists were known to have done on our
SPOT Road and really wanted no part in fouling that beautiful environment. Urination was much more common, and it was amazing
how with careful aim the deep yellow colored and steaming hole
would be cut into the ice. I always did my best to bury anything
I left and other than necessary bodily deposits, would have never
left an iota of trash. Another thing that I faced while outside was
constantly needing to forcefully blow my nose. I did this one nos-
tril at a time, while pinching off the other. I would be blowing out
mostly ice but also bloody discharge caused by the extremely dry
climate. I tended to always blow my nose at turnaround points
and sometimes, if there had not been a storm in a while, bloody
flecks littered the ice.

Interestingly, not all crewmembers shared my enthusiasm for
being outside, especially in the dark. I would sometimes make jokes
to crewmembers that being out on the smoking decks, which were

on the downwind side of the station and protected from the prevailing winds, was not really being outside. In their defense, a look outside from the station's large exterior freezer doors by someone who did not spend much time there, would not serve to entice. For some, it would probably appear hellish, with the blast of cold, the darkness, and the weird glow of the red station lights. During my first winter, some time before I would actually head outside, I would look out the station doors and marvel at the thought that I would soon be out in such conditions. During my second and third winters, I no longer felt much of anything as it had become my normal. By that time, I rarely even looked out prior to a walk. I was ready for anything. The only thing I did in advance was to try to check the wind speed, which diminished visibility as it got higher. That would help to determine the route I would take. By that time, the temperature itself meant little.

While I talked to the crew about the benefits of getting outside and experiencing something few on the planet ever would, I did not push it. There were some that were nervous, and rightfully so. My early winter outside-travel presentation, during which I stated there was "death outside that window" while pointing outside, was designed to ensure that outside travel was always taken seriously. I had one fellow, who I found out around midwinter, who had not been outside the station since the previous summer. The person that had been with him on that occasion said that he had become extremely cold at minus 10°F and wanted to go back in. During the winter he was extremely nervous about going outside, even for the winter crew photo. The photo was taken within one hundred yards of the station's main entrance. Additionally, the photo was taken after the sun had returned, and while still somewhat cold, it was mild compared to the truly cold. I noted during the photo session that his cold weather boots looked like they were brand new as they had been rarely worn. Through the winters, we had several minor cold weather injuries, mostly among crewmembers who did not spend much time outside. Usually it was to the face, which was hard to cover, and fingers, especially with photographers. Fortunately, no one lost anything except small amounts of skin.

During my first year, I had the idea that I would try to change the feeling of being cold into something else. I wanted the sensation of cold to be different for me and more applicable to what I wanted it to be. Cold is an instinctive bodily sensation and is not like a sentiment or an emotion. Being cold for most is not a welcome event. When exposed to cold the human body will do what it can to stay warm, and there will be discomfort. It does this by shivering and increasing blood circulation. Most sane people do not enjoy the feeling of being cold and instinctively do whatever they can to stay warm. Cold is a relative thing, and some people are cold at temperatures of 60°F. After my first winter, I understood what it was like to be very cold. I learned the hard way that even dressed well and almost completely covered, I would burn the bridge of my nose—and at times, other parts of my face—if the wind worked its way into gaps in my hood and balaclavas. I had experienced the intense pain of frozen hands and fingers when I took my gloves off in weather colder than minus 70°F, especially in windy conditions. The pain was intense, and it took minutes to warm up and restore feeling.

I had experienced the sensation of starting my walk and the sudden chill that went away once I hit my stride. It was somewhat like what a scuba diver with a wetsuit feels as the cool water rushes into the gap between wetsuit and skin. I wanted to change that feeling of cold to things I respected. I came up with duty, honor, and glory. Duty, for the sublime duty I owed my crew and my station. Honor, for the great honor I was experiencing in the position I was in at the South Pole. Glory, for glory to God, the creator of the universe, who gave me my life and all I had experienced. I would concentrate on those things, but it was still just cold. One day at the end of my second winter with the sun just above the horizon and the station covered in its winter frosting, it happened. I was starting my walk, and rather than the usual surge of cold, I felt a jolt like electricity that to me was the sensation of duty, honor, and glory surging through my body. It did not last long, and I never tried to repeat it. I did not need to, as once was enough.

During my walks, I would think about many things, especially if I was just walking the flag-line route out to the science buildings. On that walk, I did not really have to pay attention to anything. It was simply out about a half a mile and then back. On my grid-south walk, I had to pay more attention as I could get to a place, especially during my first year, where the lights of the station were not visible and there was no type of landmark to steer by except stars that could become obscured by clouds or ice crystals. That was my favorite walk as it was away from the station and the feeling of being so alone was wonderful. During my second year, I began using a compass and would practice finding my way back in near-whiteout conditions. I practiced with the compass some distance from the station in obscured weather. I would close my eyes and spin around trying to disorient myself. I remember a walk during my third winter when, a few miles from the station, the temperature made a rapid increase, causing ice crystals to form in the area and greatly limit visibility. If that scenario had happened during my first year, I would have been quite concerned, but at that point, it was trivial. Even with a confidence based on being outside and walking several thousand miles, I always took my walks seriously and realized just how bad things could turn.

At the South Pole I experienced a silence unlike anything I had ever known. A few miles away from the station on a still day with little to no wind I once tried to listen to my heartbeat. I was unsuccessful but was overwhelmed by the vast silence and lack of any outside motion. It was like being in a hearing-test booth, but even in those, if you listen intently, you can hear what is going on outside it. Here, there was nothing but true silence. While I found the silence interesting, I did not seek it. I loved the sound of the wind and the crack of the outside marker flags. I also filled the silent void with my own sounds, starting with practicing the Ahab iron way speech. I would at times hear music in my head and remember a walk in a darkened gale on the science flag line where I listened to my mind's own blaring stereo version of the Police's song "Roxanne," my own red light casting an eerie glow into the maelstrom. Walking grid south on a calmer day and in a reflective mood, I was able to conjure up an internal version of Simon and Garfunkel's "I Am a Rock," which I found appropriate to what I was doing at the South Pole and much of my life.

On numerous occasions, people have asked me if there was anything paranormal or weird going on at the South Pole. I know that there are some outlandish theories circulating. There are stories of secret World War II Nazi bases that still exist, and a thinking by some that the South Pole is a vortex for UFO activity and all that goes with that. *The Thing* and other science fiction movies probably help foster this. I have been outside for over four thousand miles over two summers and three winters and only once experienced anything that seemed unusual. During a full moon winter walk, I was heading grid south alongside what in summer was our skiway. About a mile from the station and to my left, I noticed what looked like shimmering shafts of light that went from the ground up into the sky. They seemed to be moving as I moved, and I could not figure out what was causing them. I had never seen anything like it in any previous walk. It was an eerie spectacle, and I kept moving, thinking about what could cause such a thing. At some point I wondered if this was a UFO-related phenomenon—which I had always laughed at. On my return, I got closer to the beams and saw the cause: ice coated communication antennas struck by and reflecting moonlight. I was surprised that I had never seen that phenomenon before and that moonlight and ice could cause it. I was also relieved to know the cause. No UFOs, and thankfully, no alien probing for me that day.

On my third winter, I distinctly remember a dark night when I had just left the station for a walk and was climbing a small hill of ice. The wind was blowing at about twenty-five knots, blasting me in the face with small ice crystals as I struggled forward. It suddenly hit me how ludicrous a "normal" person would find the situation. I started laughing. I laughed and laughed knowing no one would witness what might have been taken for insanity had anyone seen it. I did not need to be out there, but I now preferred it over just about anything. I thoroughly enjoyed the interactions I had with my winter crewmembers and always remembered the gravity of my responsibilities, but more and more I found myself withdrawing while in the station and favored the great solitude of being alone outside.

ACTIVITIES

During a South Pole winter, most of the crew works a nine-hour day six days a week. The station is set to New Zealand time to be in sync with the summer flight operations between Christchurch and McMurdo. Sunday is a day off for most. There is watch standing among power plant, communications, and a few other personnel that will make for nonstandard working schedules. The scientists have their own schedules based on the needs of the project they are assigned to. Even with what would seem to be long work weeks, there is plenty of time to do other things. When crewmembers are initially interviewed, they are asked about off-duty interests and hobbies that they have. This is a serious question, and I always listened closely to the answers. Crewmembers heavily into online computer activities would be disappointed with the minimal internet connectivity at the South Pole. The connectivity at the South Pole is sufficient to allow email, shopping, banking, and limited downloading of music and video and is limited to just a few hours a day, which could be in the middle of the night, depending on the satellites position. We ensured that all aspiring crewmembers were informed of this, to keep their electronic expectations low. I marveled later at the younger crewmembers who still carried cell phones. They worked only when the internet was up and in a very reduced capacity as there was no traditional cell phone service available. Some habits die hard, and I think some of those young crewmembers found some slight solace just having the phones with them even though they could not really use them. During the interviews, many candidates would say that they wanted to get in better shape using the gym, and some did that. Some were more realistic and stated that they liked to read and/or watch movies. Those activities are South Poles staples.

The South Pole Station has a well set up gym with treadmills, ellipticals, bikes, and free weights. There is also a full-size basketball court that can be set up for volleyball, badminton, and other sports. Each of my crews was different in the number of people who would use the weight room. Most of my first crew did not really take advantage of it until well into the season when one of them established a ninety-day fitness challenge. This was an attempt to get crewmembers in better shape for station opening and their journey home. After the long winter, several crewmembers had gained significant weight and tried to do something about that. Overall, the event was a success and was another activity that served to bond crewmembers. That crew also initiated what I hope becomes a South Pole winter tradition, the Polelympics. That winter a brilliant and off-beat electrician developed the concept and almost singlehandedly set up that grand event. While it may not have rivaled the actual Olympics in overall organization and professionalism, it had an opening ceremony, various types of competitive events, and a closing ceremony with medals.

Our amazing electrician had given much thought to the specific events, to attract and include the participation of most of the crew. In addition to physical sporting events, there were also mental contests such as a timed Rubik's Cube challenge. There was a pool tournament, which featured assigned hecklers who delighted in throwing out humorous verbal barbs that did seem to affect the play of some of the participants. I remember walking through the tournament and being stopped in my tracks by a savage barb delivered by a heckler to one of the players that concerned his recent divorce and its probable impact on his next shot. Fortunately, the barb was humorously received, although the poor, maligned fellow's next shot was terrible. While there was a level of humor that permeated most events, they could become quite serious. There was an outside sled pull that was incredibly grueling, and I remember seeing the contestants lying on the first floor of the station afterward. They were physically exhausted and barely able to move.

I also remember the fallout from one of the Polelympic events when two exceptionally well-educated crewmembers got into an extremely profane and heated altercation. This occurred when they passed each other in a station hallway. Instead of a usual greeting, one chose that moment to accuse the other of cheating during the treadmill challenge. This had supposedly occurred several weeks prior but was still a hot issue with him. The perceived cheating had knocked the instigator of the verbal conflict out of his bronze medal. I spoke to both men, and although I was somewhat amused, had to treat it as a potentially serious situation. Mere words in a hallway one day could escalate to blows on another. To most people who had never experienced a South Pole winter, an altercation over such a minor event that had occurred weeks earlier and involved a small shop-made medal would seem ludicrous. It was just another example of how the isolation and intensity of a South Pole winter could escalate passions.

I was a big proponent of winter recreation and witnessed the positive effect an active social calendar had on a crew. The activities varied from the standard winter parties, which always involved drinking, to what I favored more, events that did not. While I did support station parties and thought they could be positive events for the community, I wanted to ensure a mix of activities, including things like sports, games, crafts, travel presentations, and many other things that were positive for the mind and body. Early each winter I would ask for a volunteer and appoint a "recreation director." This person was responsible for the organization of the overall program and the coordination and scheduling of events. Event scheduling was taken quite seriously, and crewmembers tried to set up their events to not interfere with what others might have planned for that specific day and time. The recreation director would ensure the events were posted on the South Pole intranet scroll. The event's originator/sponsor would then place paper advertisements on several walls in the station's hallways where such things were posted. There was always something going on, and I really promoted my crewmembers getting out of their rooms, even to just to watch a movie. While watching a movie may not sound like much of a group activity, just sitting, talking, and interacting with crewmembers before, after, and sometimes during the movie was a good thing. I attended many movies and genuinely loved sitting with my crews for those. I did not actively participate in many other activities, but I tried to stop by and say hello while they were going on. I would also promote the events at my weekly leadership meeting and monthly All-Hands Meetings.

The interest in sporting events varied with each crew. There were always individual crewmembers interested in staying in shape who would have quite structured, scheduled, and rigorous workouts. Crewmembers had specific gym equipment that they preferred, down to which treadmill they would utilize. Someone who was erratic with their workout schedule and interfered with someone else's by utilizing "their" equipment probably felt the icy stare. My second and third crews were more active with their individual workout activities, but all three crews were incredibly involved with the team sports. During my first winter, I witnessed a sport played that I had not seen since elementary school, dodgeball. As a boy, I had relished the fierceness of trying to get kids on the other side of the gym "out" by hitting them with a hard throw from the rubber ball. It was glorious to catch an opponent's hard-thrown ball, which would then release and bring back into the game a teammate who was out. At the South Pole, grown men and women go back to their youth and play the same game, with less agility but probably more velocity on the ball. During my first winter, a crewmember was struck so hard in the face/eye that his vision was temporally affected. I hated doing it, but after considering options from deflated balls to the wearing of eye protection, I temporarily suspended the playing of the game. While dodgeball ceased to exist by that name, I learned later that a newly introduced "velocity ball," or something like that, was being played and was really the same thing. Fortunately, we had no more serious injuries. While this seems like a small thing, it is actually a good example of a real challenge with leadership. If I had done nothing and someone else was later injured, I would have been a poor leader. If I had absolutely stamped out the game and used "banning" as a regular approach to challenges, I would also have been a poor leader. Sometimes there is no clear-cut, correct solution to a problem, and you do the best you can and live with it.

Hygiene could be a challenge for athletes or others who were performing heavy work. The two, two-minute showers per week rule meant that one could not take a shower every day. It was possible to have people that smelled bad on station due to that. Fortunately, the station was kept cool, mostly in the sixties Fahrenheit range, and some people preferred their rooms to be even cooler. With that, there was less sweating. On several occasions, I was asked to talk to people after crewmembers complained to me that someone smelled bad. I was not overly bothered doing this as I may have smelled bad at times myself. I remember being on a treadmill and the person working out on the elliptical was exercising vigorously and working up a sweat. The smell was strong, and I found a reason to change machines. While we had the shower rule, I told the crew that if they needed a shower to take one. That was especially important for galley staff. There was an average number of gallons-per/person-per/week figure that was on a table on a weekly report I prepared and sent out. I quickly noted that if the number was much higher than around twenty-five or so gallons per week, people back in Denver would question it. The solution was simple. I told the facilities engineer to find a way to reduce that number. While we tried to do that by actual reductions, if that did not work, I was fine with finding any other way to reduce the number.

Some events were more cerebral, and during my first winter "Thinken and Drinken" was quite popular. On a couple of occasions, I stopped by the station library where the participants would meet. I found them engaged in deep discussions on obscure but interesting topics about which the group members could give their opinions. My observation was that the gathering started with thinken and as there was more and more drinken, there was less true thinken. It was a fun activity, and I think it helped bond crewmembers as they got to know each other better. While the Thinken and Drinken was an actual scheduled event, much discussion would occur in the lounges, especially on Saturday nights. Sometimes questions would be put to the informally gathered groups, and the subject one late night was gun control. I witnessed a very heated discussion among crewmembers on the subject. One crewmember was defending his position on why there needed to be more controls on guns, and another was arguing the exact opposite, after being robbed at gunpoint. They were both great crewmembers and although their debate was quite impassioned and probably fueled by alcohol, they were respectful and remained close friends afterward. On a more humorous note, while making my usual Saturday night station rounds, I once walked into a conversation among a group of straight men where the topic was "If you HAD to be with a man, who would it be?" The answers seemed to be primarily men who were actors that were admired by most. After pondering the question for a few seconds and not really finding an easy answer, I continued to my room and bed.

Photography and video are very popular with South Pole winter crews as crewmembers desire to document the year they spent on the ice. Many crewmembers will document their time at the South Pole with personal blogs or use other forms of social media. Prior to my first deployment, I viewed online video tours of the station to help me understand the layout. These were done so well that upon my arrival at the South Pole I could get around the station and recognized many things. One of the most popular things to photograph are the night skies and the beautiful, shimmering, green auroras. From my office, I got used to hearing the stampede of crewmembers running down the hallway to the station's main observation deck when they learned there was aurora action occurring. This would happen a few times at the start of each winter, when just a smudge of green might be showing in the night sky. The interest would usually taper off as the auroras became more common. Some of the crewmembers arrived at the station with amateur photographic skill, and by the time of their departures, they were quite accomplished, with many stunning images. Our winter crew photos, which will hang in the South Pole station hallways for as long as we have a station there, are testimonials to the superb vision and photographic skills of Hunter, Viktor, Yuya, and all who assisted.

On my third winter, the Explorers Club, of which I was a member, asked members to document with a one-minute video what they were doing as the COVID virus had severely impacted travel and most types of exploration. The video would be distributed among members and was called "A Minute at the Mic." Having no video skills, I asked Danny, our steward, to film it for me. I had noted that he was becoming quite adept with his outside photography. My minute turned out to be only forty seconds in length and consisted of a greeting and my recital of the Ahab iron way lines. I had practiced those lines while out alone on the ice, and they were by that time emblazoned in my being. Because of that, I was able to do it in one take. This was important because, although I was wearing my wolfskin anorak, I was wearing my station trousers and only one balaclava as I wanted my face exposed. We filmed on the station's observation deck, with temperatures in the minus eighties Fahrenheit, and it was windy.

The result was very striking, and during the video, I had made a comment about our station being a ship atop moving ice that was flowing to the Southern Ocean. Because of that, I was asked by the Explorers Club to participate in "World Oceans Week," with a longer video. Once again, Danny and I set up outside on the observation deck and, thinking I could do it in one take, I again dressed lightly. Unfortunately, this video was more complex, with more dialogue and a specific order. The temperature was around minus 83°F with a strong wind that made the windchill around minus 130°F. By then I was a believer in the effect of windchill, and the experience that day out on the observation deck really cemented that. On my first attempt to speak, I forgot my lines as I was in shock over the icy blast hitting my unprotected face. The next attempt ended abruptly after the extreme cold knocked out the batteries in our cameras light and we stood in darkness. It took several takes and some editing to get a final product. Looking closely at that video, I could see where ice was forming on the side of my nose. What the video doesn't show is the freezing inside my nose that hurt for several days after. All in all, Danny and several other crewmembers that assisted put together a beautiful, informative final product. Sometime later, at an All-Hands Meeting, I played for the crew another version of that video, a little different from what was submitted to the Explorers Club. In that version, Danny had included the outtakes. With my foul verbal reactions to forgetting my lines and the face-freezing process, it was not something to show children.

Not all activities were planned, and many impromptu events occurred. My last crew, which had a core group of younger members, was especially active with things they considered fun as their world was so impacted by quiet cell phones. I was somewhat surprised early in the winter when I was asked if they could play hide-and-seek in the station. I told them they could, but it needed to be done safely and certain vital areas such as the emergency power plant, EPP, were off limits. The EPP has several exposed gauges that could easily be struck by something and broken. It also has a CO_2 fire suppression system that should not be unnecessarily activated. In the dark, antics could possibly do both. While I did not participate, I did make sure the crewmembers were all present and accounted for the next morning. Another oddity of that crew was that some of them would get together to watch movies and cartoons that were made for ten-year-olds. While I found it somewhat immature, it did no harm and was a lot better than if they had developed a passion for snuff films. With that crew, I was not surprised when the idea for a "slumber party" was floated among them, but for some reason, the event did not develop enough interest. Although young and at times immature, there was great turnout for my early winter showings of the Antarctic films during Adventure Movie Night with Wayne, AMNWW. They really did care about the history of the South Pole, and I was incredibly happy to see that. Another great virtue they all seemed to possess was the ability to put the audio and video cables back in their correct locations and place the proper settings back on the equipment prior to my AMNWW presentations. As they were the ones that had temporarily reconfigured the system to play computer games, on more than one occasion I had to do a radio call to get one of them to put it back.

Religion was practiced by some, and during my first winter, we had an ordained minister on our crew. He did a great job of conducting a nondenominational Sunday service, being available if people wanted to talk, and not pushing religion in people's faces. On that crew, there was also a meeting of atheists that watched religious debate videos and discussed topics related to atheism. I stopped by a couple of times and noted it was an all-male group with an edge against organized religion. After listening to some of their angst, I really thought they just needed girlfriends. Through the three winters, there were Christians of various levels of conviction, a devout Jew, a devout Mormon, and atheists on the crew. Most members of the crew kept any religious convictions to themselves. For the most part, they coexisted quite well. One event that I never really understood but tried to remain silent on was Christmas in July. Through the years, I had witnessed attempts by retailers to try to promote this to sell things, and while not finding it offensive, I was no supporter. At the South Pole, the real Christmas religious holiday is not really supported with the vigor you might find in other segments of society. The religious nature is mostly ignored, and the biggest celebration is the beautiful dinner the galley puts on. With that, I was mystified that some of the crewmembers would want to put the fake Christmas tree back up, decorate the galley, and hold any type of mock holiday in July. It took me until my third winter to see the positive nature of the occasion and that some crewmembers really enjoyed it. For the most part, I just wanted to be assured they would put the decorations back in storage after it was over.

During my third winter, which was back-to-back, I arrived back on station after my PQ in Christchurch and a few weeks at home. Upon my arrival, I found an interesting item in my office, a seven-foot-tall, wooden cross. I was told that Italian tourists had brought it in the summer, along with a rosary that had been blessed by the Pope. I was told it was to be placed at the Geographical South Pole at a specific time on Easter Sunday. Their plan was to simultaneously plant another cross at the Geographical North Pole. I was surprised by the request but awaited further instructions. The instructions never came. I assumed the COVID crisis ruined their North Pole venture. Hearing nothing from the people who had brought that cross at great expense to the South Pole, I debated what to do and then decided. On Easter Sunday of 2020, I, a non-Catholic, carried that cross and the Pope-blessed rosary out to the Geographical South Pole. I planted it near the pole-marker sign and took a photo. It was not especially cold, but it was very dark, with some wind and blowing ice crystals in the air that obscured visibility. I did not claim the South Pole for the divine Papa in Rome, but I felt good that on Easter Sunday, while the COVID virus ravaged the world, that beautiful cross and all it means stood at the Geographical South Pole.

Station meetings were another example of scheduled activities. Weekly I conducted what I termed a "leadership meeting" with station personnel who functioned in leadership roles. We met on Saturday mornings in our large conference room. I would start by discussing pertinent station topics, such as feedback on an emergency drill; pass on information that I may have been given from the Denver office; and ask the members of the group questions. We would then go person-to-person, when each department head could give the group pertinent information, such as internet availability, galley happenings, and any noted safety concerns. I kept the tone of the meetings positive as I had learned years earlier that in some organizations there were people who thrived on the opportunity to criticize or denigrate others, especially in a public forum. With this type of person, meetings can quickly devolve into a negative mess. In general, South Pole winter crewmembers were not of this type, but I was always watchful and wanted to maintain an atmosphere of respect for all. I would usually humorously end the meetings with an advertisement for my upcoming adventure movie night.

Once a month at the conclusion of a regular weekly leadership meeting, I would keep a small group of these leaders in the room for what was known as the "collegial committee." The committee's existence and purpose were spelled out in a procedure and had been mandated after a disastrous winter where some crewmembers had run amok. While I do not think the existence of this committee would have made the slightest difference that winter, I supported the concept. The purpose of this committee was to discuss station occurrences of a sensitive and sometimes personal nature. At this meeting, such things as a crewmember that was having a serious off-ice family situation or one that seemed to be developing a problem with alcohol could be topics of discussion. The group then discussed how to deal with the situation. The most common issues were probably criticisms of food. All three of my crews had people that would do this, even though I and most others considered the food quite good. There were always a few that would complain. Some of the criticisms had merit, as maybe we were serving too much or too little of a certain dish. Others were just people's personal preferences, which can radically vary. With the criticisms, I was more concerned with the morale of our galley staff, who tried so hard and with such pride to cook things we would enjoy and at times would hear negative comments. I chose the members of the collegial committee quite carefully. I wanted members that I thought genuinely cared about people and their problems and who I could count on for total confidentiality. The mission of the group as I saw it was to help alleviate a problem before it escalated. I could not be everywhere and see everything and really needed to gather information through others. Serious issues were discussed, and I very much appreciated the committee members opinions and willingness to help.

The All-Hands Meetings were something that I particularly enjoyed and found useful. These were scheduled on the first Saturday of the month. They were opened by a short presentation from the safety engineer, who would discuss issues such as electrical safety, housekeeping, or safety issues that were timely and pertinent. If a crewmember had suffered a recent injury, he would discuss that and what we needed to do to prevent another. After the safety presentation was completed, I would start my presentation. I would go through a list of topics such as information I may have received from the Denver office, station cleanliness, upcoming events, and cautions, if I saw the need for such. I was always concerned about social events with alcohol involved and wanted the crew to be ever vigilant. My crews were quite aware that the use of alcohol was a privilege we enjoyed, and they did not want to be the crew that was responsible for it being cut off. There had been discussions in the past about making the station alcohol free or "dry." This would have been consistent with places such as the North Slope of Alaska. Although I did not drink during regular winter events, I wanted my crews to be able to and encouraged them to use common sense and to take care of each other. I would also discuss future items, such as station opening, and would be quite clear with guidance on how we would deal with each situation, but I readily called on crewmembers to give their opinions.

My style of leadership was not anywhere close to a democracy, as some WSMs favored, it was more a benevolent dictatorship or monarchy, with a leader who was very willing to listen to and consider crewmembers' thoughts and opinions. I knew that most people preferred a strong leader but not a tyrant, bully, or micromanager. If someone had a better idea or path forward, I was always ready to embrace and promote it. At the meetings I would use my type of dry and at times slightly outrageous humor. As my life was outrageous, it could be difficult for crewmembers to know when I was joking and when I was serious. What was normal for me was not for most. The crews would initially take a while to understand it, but they finally would. At these meetings, I enjoyed showcasing people who had done a great job with something. There were many great jobs done by my crewmembers and I wanted them to be recognized, especially in front of a group. After I was finished, I went around the room and the crew's leaders had a chance to pass along any pertinent information to those assembled. Common topics were things such as galley changes, internet usage, and an upcoming audit. I wanted the leaders to be leaders in the eyes of the crew. I would then let anyone that needed to make an announcement do that, and that might include a lost item, upcoming social event, or the need for volunteers to help with a task. I tried to keep the tone of the meetings positive and on track, and I wanted to promote the idea of a crew, our crew—the only crew that would ever be for that specific year.

During my third winter an interesting thing occurred at an All-Hands Meeting. Prior to the meeting I had been asked by a crewmember if I was okay with something funny being done over things I had said. This puzzled me, but I told her I was fine with whatever it was. The meeting commenced, and after I had finished with what I needed to say, I told the assembled group that now was fine for whatever funny thing they had in mind. It was a little confusing, but I soon figured out that the "funny" event had already begun. I saw that many of the young devils had bingo-type cards with preprinted phrases and topics contained in squares. Printed in the small squares were phrases I had used at previous All-Hands Meetings. Each crewmember was anxiously awaiting me to a repeat something so they could make an X through it on their bingo card. I looked at a card and there were such gems as "goddamn," a word I was known to utilize rather frequently. It also included "Are you guys happy to be here?," which I know I had asked on many occasions and always hoped that they were. One block said that I would do a commercial for my adventure movie night, which in fact, I always did. One block said I would "repeat a story," which I knew I did and for a reason. There was block that said "young drunks" and one that said "old drunks." This came from labels I had given the two different types of South Pole station drunks, which, fortunately, were mostly confined to Saturday nights. I had explained at previous All-Hands Meetings that the "old drunks" did not worry me as much as the "young drunks." This was because the "old drunks" tended to drink beer, sit in the lounge, and discuss topics they had discussed the previous Saturday night and had probably forgotten about. They were quite experienced in their drunkenness, after years of practice, and mostly went to bed early. The "young drunks" worried me more. They were less experienced in the art of drunkenness. Being young, they were more active during their drunkenness and more likely to think and do stupid things that could result in problems. Of primary concern were the least experienced drunkards who had not yet developed any kind of sense for limits. They could be enticed to drink too much of some shitty, spiked fruit punch. This would make for interesting vomit later. I watched out for those. All in all, I found the Wayne bingo terrifically clever, amusing, and a befit to the South Pole crew.

In defense of my repeating of stories, I knew I did that, and I did it on purpose. It was from the teachers' "rule of three." The rule of three states that you should give students information three times before you expect them to use it. Maybe for me it was ten times, but there were concepts I really wanted my crews to understand and embrace. A phrase I developed and used for every crew was this: "You are the 2017 (or 2019, and finally, the 2020) South Pole winter crew. You are the ONLY 2017 (etc.) crew there will ever be." By this phrase I wanted my crew to understand the significance and permanence of their experience during their winter at the South Pole. They were now a member of an exclusive group, a specific year's South Pole winter crew that could never be repeated. What they did counted and counted for eternity. They would be photographed with their crew, and that photo would be placed on the long station hallway. They and their crew would become immortal. I repeated that phrase for a purpose. I wanted it burned into them.

During the long winter, the South Pole's emergency response team, ERT, must be able to respond to any kind of emergency. Of primary importance is the fire team. The fire team must have the training, equipment, and leadership to put out any fire that could possibly occur on station. For my first year, I was quite nervous about this. For years, working at remote sites around the world, I had the luxury of having professional fire departments with real fire chiefs reporting to me. These real firemen were extremely professionally qualified, well trained, and quite capable. Their qualifications were rigorous and mandated by contractual requirements. At remote sites, I was the contract's senior person and leader of the Incident Command System, ICS. That position is responsible for the coordination of resources during an emergency. While I occupied that position, when it came to fires, I always deferred to the vast knowledge and skills of my professional firefighting staff. Now I was in a place far more remote with no hope of any kind of assistance, and there were no professional firefighters. At the South Pole, we had a volunteer fire brigade with one week of training and some on-station practice. Knowing our ERT limitations, I stressed fire prevention over being able to put out a fire. Even doing that, we had incidents.

One of the first things a newly arrived South Pole winter crew does is an ERT turnover drill with the departing winter crew. There is a desire by the outgoing crew to do this as soon as possible as they want to be relieved after having the duty for a year. There can also be a sense of nervousness by the incoming crew as most have only had a week of fire school and are now expected to show their proficiency and assume an important duty. Understanding this, the drill is quite simple and is nothing but a start and a confidence builder for the incoming winter crew. Many times, the initial drill is fire in a lounge and the extraction of a victim. During my first winter, my newly minted fire team did a great job extinguishing the mock fire, locating and then extracting the victim. The only real complaint came from the victim, who was from the outgoing winter crew and whose hair was quite long after a year with no haircut. I thought he was exceptionally good natured about the fact that while someone was trying to pull him out of the fire, someone else was standing on his hair.

After the turnover drill, the incoming crew will need to continue to train and drill to improve their proficiency and ability to respond. Drills are preplanned events and vary in level of challenge. They might be something simple, like a report that smoke has been observed in a room, or more sophisticated, with a complex scenario that involves mass casualties. At the completion of a drill there should be a debrief that includes all the participants. During the debrief, the drill is discussed in detail, including positive things that were observed and anything negative. People being people, the negative sometimes seem to gather more attention. The main goal of the drill and the debrief is to gather valuable lessons learned that can be used to improve the response process.

During the summer, emergencies are handled by a mixture of summer people, most of which have not had formal ERT training, and the trained winter crewmembers. This mixture is necessary to fill all required emergency positions, as several of the winter crewmembers are winter only and will not arrive until just before the station closes in February. This means the winter crewmembers on the summer ERT will have had at least some formal emergency training, while the summer members may have only an interest. I learned to dislike this mix as there were some summer people who would have the interest and motivation but little ability and there was no real long-term benefit to their being on the team as they would be departing at the end of summer. This could create problems during drills, training, and any actual emergencies.

My first few experiences with South Pole drills were not the most positive events. I believe in the concept of conducting drills and would tell my crew an old Marine Corps training adage: "Sweat now in peacetime rather than bleed in war." The problem was that it seemed like we would "bleed" during the summertime emergency drills. The drills could quickly turn into disorganized messes as frustrated crewmembers turned on each other. It was more than apparent that with our small amount of training many mistakes would be made. We were not a professional emergency response organization but a group of volunteers doing the best we could with what we had. The debrief after a drill could be a bloody affair as some people used the forum to point out perceived deficiencies in others and their actions. This would usually rankle the person being pointed out and did nothing good for the process. At the end of my first summer, I must say, I was happy when I thanked the summer ERT volunteers for what they had done and watched them depart. I learned valuable lessons from that experience and during my next summer was more reticent about placing summer people on the ERT. I also very quickly and forcefully, during the debriefs, intervened and stopped any kind of negativity aimed at specific crewmembers. I only had to do that a time or two before a tone was set that such actions against others would not be tolerated.

The South Pole Station and its associated buildings have a sophisticated fire alarm system that, when fully operational (and its information understood by the ERT), provides a great deal of fire protection. When fully operational, the system will give an audible alarm that will be heard throughout the station. This system will also automatically close fire doors and emergency lights will flash. In the communications center and the hallway walls on the station's first floor, there are map-like panels that will indicate the exact location of the fire. Every berthing room has its own sensor and alarm. For the system to be and stay fully operational, it requires constant attention and maintenance from a fully qualified and certified fire alarm technician. That position is difficult to fill, and for my first winter, we did not fill it and had no technician. We tried to keep the system operational by utilizing our extremely knowledgeable facilities personnel. We gathered technical information from off-ice experts by email and phone. This did not work, and toward the end of that winter, I had little faith in the system. We were getting false alarms, and I had reason to believe some of the system would not function if we had a real fire. I contemplated having a fire watch walk the station at night but in the end, just ensured the crew was practicing fire prevention and was always on the alert for a potential fire. We had personnel who worked the night shift and could provide some coverage, so at least a few people were always awake. For my second and third winters, we had an alarm technician. The only issues then were complaints from crewmembers who were annoyed by the constant testing. I was not annoyed and slept better knowing we had that protection.

One event that caught my attention the first winter was an audible alarm alerting us that we had a fire in our power plant. This was on a Saturday night. I considered this the worst possible time as it was the only night the crew had before a day off. It was not a good time for an emergency response as many crewmembers were socializing and drinking. It occurred in the middle of Adventure Movie Night with Wayne. *The Sand Pebbles* was being shown when the alarm went off. The movie was suddenly stopped, and people ran to their emergency positions. While we had a duty roster, which kept a four-person fire team from drinking, there was always concern that someone would ignore that. With such a small group on a no-drinking watch, I was always aware that in the event of a major emergency, I might not have as many fully capable fire-team members as needed. Because of this, I made a point of having the fire-team leads understand the need to ensure that all their team members were fit for duty at any emergency event. To my knowledge there was never a fitness-for-duty incident. I assume the team leaders handled any that occurred. Fortunately, in the case of the fire in the power plant, the power plant lead was on watch and had used a portable CO_2 extinguisher to put out the fire on the alternator of one of the generators. The fact that it was near the unit's fuel line gave us something to think about later. It was not a major emergency but a great reminder for me that I was always on duty. I experienced several minor fire incidents during my three winters.

Another small fire occurred in a fan room when a motor burned. In that case it was mostly smoke, but a lesson learned was the fact that the windings on a motor can be coated in substances that when burning may give off toxic gases. The smoke in the area was extremely acrid and our maintenance personnel ignored it to start repairs. We discussed this afterward, as they should not have been breathing it and should not have had access to the room until it had been thoroughly ventilated with our emergency exhaust fans. The same type of event occurred during my third winter. This time, ERT members cordoned off the room, wore SCBAS for entry, and set up the ventilation fans.

One event that was more amusing happened early one Sunday morning just after midnight. Someone had inadvertently left a plastic water bottle on the heating unit in the station's sauna. They turned the sauna's heat on and did not notice the bottle sitting there. They then left to let the sauna heat up and went back to their room to change clothes. Upon their return, the sauna was filled with an acrid smoke from the melted plastic bottle. I soon after arrived at the scene along with several others who smelled the smoke. Not wanting to initiate a station emergency at that hour, we tried to evacuate the smoke and let the crew sleep. We set up a large fan, opened doors, and made a path to the vertical tower to blow the smoke out of the station. Unfortunately, the smoke was so thick it made its way down a hallway that we did not want it to, and smoke detectors were activated. Even with a quick silencing of the system, I marveled as I immediately saw newly awakened fire-team members running down a hallway to get to their bunker gear. I was able to stop them and terminate the emergency but was extremely impressed by how they responded.

My next two years with the ERT were a lot easier as I had learned so many valuable lessons during the first year. I was able to sleep better at night knowing our capabilities. I was fortunate to have fire-team leaders who took the assignment quite seriously, and I knew they and their team would do their best in the event of a fire. What concerned me most was keeping up the level of enthusiasm to be on the fire team. Fire-team members were required to participate in weekly training and the duty roster prevented them from drinking on certain nights. An additional requirement was that all members must be clean-shaven. This was so they could get a proper seal with their self-contained breathing apparatus, SCBA. This became a challenge in retaining fire-team members as some men undergoing a South Pole winter prefer to grow beards. The inability to safely wear a SCBA with a beard would disqualify them from the fire team. The beards mark their time, and by the end of winter, they can look quite unkempt and shaggy. That is how they want to look and at the end of a winter, will return to the world that way.

The fire team was voluntary and there were few positives to being on it other than a desire to serve the community and the camaraderie. During my first winter, we lost people on the fire team who asked to be taken off for various reasons, such as their shift-work schedule that made volunteering difficult or the growing of a beard. I could not force someone to wear a SCBA and fight a fire. During my first winter, I gained valuable experience and insight into things I could do better. I became more proficient with my incident commander duties and had a much better understanding of what the best response should be to different types of emergencies. I became better at placing the best possible people as team leaders and gave them all the support I could. For the next two winters, I was extremely clear from the start that I needed people who would stay on the team and not ask later to be removed. This paid dividends, and after struggling and learning alongside my first crew, I was immensely proud of the dedication and enthusiasm shown by my last two. I had no sleepless nights due to them.

I was always genuinely concerned about the prospect of a serious medical emergency during winter. At the South Pole during the winter, a medical emergency becomes extremely serious as there is no way to evacuate a person. In 2016 there was a historic midwinter evacuation of two ill winterover crewmembers. I was in the Denver office at that time preparing for my first winter, and I witnessed the planning of the event and its outcome. I think the planning and execution by all involved was superb. I also knew enough about the event and how it came together that while at the South Pole I never looked at it as something that could happen so successfully again. For the 2016 midwinter evacuation, the stars truly aligned with the low winds, moon, and an acceptable winter temperature, which would not freeze the aviation fuel. The flight crew of that Twin Otter aircraft was exceptional and was able to fly it heavily laden in darkness, cold, and at altitude back to safety. The station personnel performed their duties exceptionally well and developed a very professionally written procedure to be used in the future. I dreaded a situation that would cause the need to repeat such an event. I made it clear to my crews that if it were I that needed evacuation, I would stay and not risk the safety of flight crews. I understand that in the distant past, crewmembers at the South Pole signed a document in which they acknowledged that if they were seriously hurt or ill, there was no guarantee they would be rescued. I think such a document has merit as I do not believe it is fair to risk the lives of others in such a situation. There is a station procedure for how to store a body during the winter. It has been used before.

The station's medical response team is headed by a single doctor and a physician's assistant and/or a nurse. The medical response team practices throughout the winter for medical emergencies. I found that they could quite effectively deal with minor incidences and would just do their best with anything major. The reason for a stringent PQ process was to try to eliminate medical situations before they occurred, by hiring healthy individuals. During my three winters, there were only a couple of occurrences where crewmembers were ill enough to be confined to the clinic. I remember one patient confined to the clinic who asked me quite emotionally if there was going to be an air evacuation. I told him a definitive no. While he was quite worried about his health, I do not think he fully grasped the magnitude and risks associated with such an evacuation. Fortunately, through the care and expertise of our medical staff, the crewmember recovered and returned to full duty. The medical-emergency team had weekly training, including topics such as patient transport, stop the bleeding, CPR, and other things. My third winter really stood out as I saw an amazing amount of enthusiasm exhibited by the medical team during their training. I give full credit for that enthusiasm to them having an amazing and caring doctor leading them. Overall, I found that by the end of winter, they were always a very qualified and cohesive team. I was glad we never experienced a severe enough medical emergency to need their expertise.

Safety at the South Pole is of paramount importance. Because I had a professional safety background, I tried hard to stay on alert for things that I saw might lead to an accident. The main thing I stressed to the crews was that no one would ever be pushed to a pace that would create an unsafe condition. There were times, such as after a power outage, where expedience was necessary, but for the most part, we had time to accomplish tasks safely. There was a safety engineer on each crew, and their selection was especially important. Working remote sites around the world, I had experienced the types of safety professionals I did not want on my South Pole crews. The first type was the amateurish "ticket writer." These were usually people who had risen to work in the safety field as a collateral duty from their primary assignment in a tool room or other mundane duties. They were usually without a college degree and lacked meaningful certifications. These people enjoyed the power they had and would revel in stopping a job. Jobs can and should be stopped should an unsafe condition exist. The job of a real safety professional is to find ways to work safely, not just stop jobs after unsafe acts are found to be occurring. The second type was the desk/computer-bound type that had long ago lost the desire to actively participate in the work itself and preferred to stay seated and away from it. Many of this type exist, especially in the area of safety management. Someone who had worked as a safety manager in previous jobs may have had a staff to do the actual work. The South Pole safety engineer was a solo position, and the person needed to be ready and able to do anything required to provide a safe work environment for the crew. It was then up to the crew to work safely. On my three winter crews, the safety engineers were active participants in the work process. They were integrated into the facilities department and were key to the excellent safety record we maintained.

My professional safety experience included a tragedy that occurred early in my career. Working out on a remote island, I was the environmental/safety manager for a large company that employed foreign nationals from the Philippines and other countries. There was a project on the far and more remote side of the island to dig a trench, which fiber-optic cables were then to be placed in. Two men were assigned, Pedro, a young laborer, and an equipment operator, who would operate the trenching machine. I visited the site daily and the two were enjoying the job. One day I discovered them both happily fishing in the lagoon during a break. All seemed well, but it wasn't, as one day Pedro was killed when he fell into the trench and was struck by the ferocious cutting blade of the trencher—a trencher designed to cut through coral. Pedro was married and his young wife pregnant with his child. Now he was dead. He had died horrifically, in a trench. I don't know if there was anything better or different I could have done to prevent his death, but it hit me very hard. I was given, for my files, pictures of Pedro lying dead on the medical gurney with terrible injuries and lifeless eyes. I kept a copy of those photos, and every year on the day of the accident, I would look at them, remember that day, and try to do better. Pedro died on Valentine's Day. I now, and for many years, have known that day only as the day Pedro died. I wanted no occurrence like that at the South Pole.

Another type of emergency that happened much more often than a fire or medical emergency was loss of power from the power plant. When this happened, it was quite apparent as the regular station lights went out and the emergency lighting went into automatic operation. There were audible alarms that told the facilities department that power had been lost to critical systems. Of primary importance was the drinking water. Upon loss of power, the station maintenance technicians damn near ran the several-hundred-yard path to the Rodwell building. Time was critical because if the water line contained within the ice tunnels fully froze, there was no way to unfreeze it. The maintenance specialists needed to ensure that the water pump was operational and was pumping water once power was restored. If we had ever had an outage so severe that the water piping was frozen beyond our ability to unfreeze it, we would spend the rest of the winter utilizing our snow melter. While we had an inexhaustible supply of ice to melt, the snow melter required constant filling of ice in its hopper and would supply the station water at a reduced rate. That would mean strict water rationing. This had happened in at least one previous year and had caused great dissent among the crew.

During the summer prior to my second winter, a horrific event occurred at the generator building that powered the radio transmitter on Mount Newall near the Dry Valleys. Two workers from McMurdo Station died when a fire suppression system activated while they were working on it. The release of CO_2 displaced the oxygen, killing one at the scene, the other dying a short time later. This accident sent shock waves through the USAP, which was then focused on the ASC as the two men were contract workers. Procedures concerning fire suppression systems were scrutinized, updated, and an investigation was conducted. I never saw the investigation and imagine the tragic event turned into a legal nightmare. The USAP as an organization always stressed safety, but the tragic event really caused a major reevaluation. Even though the incident occurred near McMurdo with McMurdo personnel, South Pole really felt the pressure to have no such incident. Then the unthinkable happened. We did.

Our incident resulted in no deaths or injuries. It occurred at the very end of summer, when a newly hired winter alarm technician pulled the wrong pull station during routine testing. He was with a Denver-based coworker performing fire alarm testing, which had nothing to do with the fire suppression system. He thought what he pulled would send a fire alarm signal, but instead, it flooded the communication dome with CO_2. No one was injured because he heard the alarm and quickly exited the dome. It was a sickening situation as it put those of us at the South Pole on the radar for unsafe practices. South Pole Station management including myself did our best to defend the worker, as there were some anomalies with that specific CO_2 pull station. Even though the pull station itself was clearly labeled CO_2, the conduit that contained the wiring to it was labeled "Fire Alarm." That served as part of an explanation as to why he had pulled it. Due to the station's focused and strong defense, the worker was not terminated, but all such work immediately stopped. We went into that winter in a hole of sorts, with much of our work now requiring extensive preplanning and numerous signatures. Some work, such as anything to do with the fire alarm system, was stopped for months. We finally emerged from that hole after months of having no further issues.

WINTER ISSUES

During my three winters, I worked hard to try to prevent negative events from occurring that involved crewmembers. I always felt that prevention was better than simply reacting to incidents afterward. It was like fire prevention, in which it is much better to practice preventative measures than to have to fight a fire. I had seen many negative interactions during my years working at remote sites around the world. These ranged from simple verbal disagreements to bloody fights. I knew that the genesis of these occurrences could be seemingly innocuous events that spiraled out of control if not quickly checked. I also knew that I had not "seen it all," which rang true with my experience at the South Pole. At the South Pole, especially in a winter, situations can be compressed and magnified. Someone's simple quirks might enrage a fellow crewmember. My boss used to use an example of someone being enraged over the way another ate his soup. Unfortunately, in this world there are people who are "button pushers," who revel in making someone else miserable. There were examples of this from the past when people had been pushed by someone to a boiling point and mayhem erupted. At a Russian Antarctic coastal station, an engineer stabbed an electrician over a seemingly trivial matter. I learned to be constantly on watch for things that could boil over. One example of this was a carpenter who, in the middle of winter, sent out a terse email to the whole crew that said someone had taken his shop clock and he wanted it back. While this might seem like a small thing, it had the potential of turning into something much larger. In this case, a clock was returned, not the same clock, but one that the carpenter felt was a reasonable substitute for the missing one. What he probably never knew was that his boss, sensing what could happen next, quickly found another clock of about the same type and replaced the missing one. Crisis averted!

"Ice wife" is the term for a wife that one may have or be only while at the South Pole. This is usually in no way a permanent relationship. A "tarmac divorce" usually follows. The tarmac being the runway surface at the Christchurch airport where winterovers emerge from Antarctica. The ratio of men to women at the South Pole is quite unnatural, with far more men than women. Women who in normal society would be considered of average attractiveness become instant beauty queens and can obtain a temporary mate rather easily. Overzealous males can then do idiotic things to try to attract the few available females. During the South Pole summer, there were plenty of quick hookups and breakups. The winter was a different situation. There are horrific stories of breakups that have occurred during winter, one being the case of a married couple in which the wife decided to split and quickly found a new suitor. She then moved into her new mate's room, much to her husband's horror. Breakups are common occurrences. Besides the turmoil the couple faces, the crew can at times pick sides, and the breakup then becomes something even worse. Fortunately, during my winters, I only had to deal with one breakup of a couple on a crew. In that case, they were both excellent crewmembers but probably not the best match. Once the mismatch became apparent, one of them initiated a midwinter breakup. There were then hurt feelings by the other, who had a hard time accepting the change. They were both superb people, who I cared for, and I talked to both, trying to do what I could to get them through the winter. This was one of the cases when I broke character and gave examples of my own silly love life from many years prior. I wanted to defuse the situation, and most importantly, I wanted to try to control things so as not to affect the rest of the crew. Love truly blooms at the South Pole and can wither just as quickly.

As a South Pole winter begins, there are always crewmembers who step up and get immediately involved in volunteer activities. These activities include fueling the last transiting aircrafts, working in the greenhouse, taking station fuel readings, helping in the dish-pit, and other things. Initially, there is always great enthusiasm and people come forward to try and do their part by volunteering for many things. Their enthusiasm sometimes wanes after they realize that many on the crew have no intention of helping if they do not have to. Resentment can build from this. I noted that during my first winter and saw the negative results from it. For my second winter, I encouraged overenthusiastic crewmembers to somewhat temper enthusiasm and try to pace themselves through the long winter. A common occurrence during a South Pole winter is the development of a crewmember who believes they are the hardest working person on the station. A martyr syndrome develops, and the self-perceived hardest working would then spend time watch-ing and noting that others did less than they did. I knew of a text-book case that involved a martyr from another winter crew who worked in the power plant. He was extremely vocal about his value to the crew, although his boss and other crewmembers thought lit-tle of him and his abilities. He was placed on the first aircraft out at the end of that winter and was in no way missed.

I was extremely fortunate that during my three winters at the South Pole I never faced any severe personnel issues. In my career at remote sites around the world, I had the luxury of being able to terminate employments and send a troubled employee home if it came to that. Sometimes, out on the Alaskan Aleutian Islands, with their horrific weather, it would take a while, but eventually the day would come when the problem was loaded on an aircraft and followed by a sigh of relief by many. At the South Pole during winter, that option did not exist. I think that recruiters and hiring managers did a great job of vetting most prospective crewmembers. The face-to-face interview eliminated most of the truly troubled. The teambuilding and emergency response training helped eliminate others. Even with that process, there was a hole of sorts that from time to time allowed someone to winter that should not have. This occurred with last-minute hires who did not get the same scrutiny and might arrive on one of the summer's last flights. We might have only days, and sometimes not even that long, to ascertain the person's fitness for winter. One of the worst South Pole winters on record was heavily impacted by a last-minute hire who did not receive the required scrutiny and showed his true colors after the last aircraft departed. We always viewed the last-minute, late-arriving crewmembers with some trepidation, but fortunately, during my three winters, they had not been a problem.

No matter how we screened, every winter some of the crewmembers had problems. Some of these problems were difficult or impossible to predict and were caused by events that occurred off-ice. Off-ice occurrences can have very severe impacts on South Pole winterovers. A death or severe illness in the family or an unexpected divorce or breakup were among the worst things that could happen. I had crewmembers who experienced those things, and when such an event occurred, the community usually knew about it. There are few if any secrets at the South Pole. We encouraged community members to rise to these challenges and provide support and assistance as best they could. I was extremely proud of my crews for stepping up and rendering much needed support during these times. I was always available for anyone that needed to talk to me, but I found it inspiring when I saw the crew taking care of each other.

During my three South Pole winters, I experienced only a small handful of troubled crewmembers that, in retrospect, should not have been hired. One of the challenges with hiring the winter crew is selecting crewmembers who have just completed a winter and want to return for another. This is known as doing a back-to-back winter. One would think the process easy as you have a year's worth of information on the person and how they performed, provided by the outgoing WSM and the persons direct supervision. The reality is, there were people that might have been okay on one crew but not another. Most back-to-back, prospective crewmembers probably thought the interview they received from me after their winter was a mere formality. Some were then shocked to learn that they were not selected for my crew. My feeling was that a person who was mediocre and had merely survived a winter was not good enough. One person who stood out was a genuinely nice person who I personally liked and wanted to select. That person had experienced several lapses in judgment during the previous winter. These lapses in judgement resulted in damage to the station. One specific act could have been catastrophic. During the interview, I was not assured that it would never happen again, and we made the decision to not rehire for that winter.

Something I learned during my second winter was that even a person who had wintered before was never assured of a good second winter. I saw that many crewmembers seemed to remember their first winter as a magical time. So magical that nothing could ever compare to it. With that thinking, the next winter could never measure up to the first. If all things were honestly considered, those first winters may not have been as magical as remembered. In fact, I never met a crewmember who stated that their most recent winter was better than their first. Some crewmembers who had wintered previously had a hard time adjusting to the fact that it was now a new winter and a new crew. I witnessed several crewmembers during a winter begin to cry at the thought of past friends from a previous winter who were not with them now. One night, while making my station rounds, I heard a statement from a crewmember that so-and-so was not there and he missed him. I pointed to another crewmember sitting nearby and said, "But he's here!" I think just by coincidence the two did become fast friends, and I laughed out loud when I heard a couple of months later that they wanted to leave together on the same flight at the end of winter. It is a South Pole truism that some veterans constantly bombard new crewmembers who are on their first winter with statements like "in such and such year this or that happened, and it was a much better year and could never be compared to." These statements get old.

At the start of one winter, while speaking to my new crew, I criticized some aspects of the crew that had just departed. Noting their winter, I saw some silliness creeping into the station that I did not want in my crew and upcoming winter. Several of that crew were now on my crew as they were to do a back-to-back winter. I wanted to catch their attention. They had been a good crew, but their leadership allowed a few things to occur that I personally found distasteful and would have handled differently. These events included a couple of women who seemed to enjoy turning men against each other, a jilted lover screaming in the hallway, an idiot who for a prank thought it cute to pound on people's berthing doors, and for some reason, the use of a fat-tire bicycle to ride down the station's hallways. As they were preparing to depart at the end of their winter and my crew was arriving, I personally witnessed one of them in hysterics in the station's hallway over an aircraft that did not arrive. Aircraft not arriving at the South Pole is a common occurrence and people must learn to deal with it. She received great attention during her display from her crewmembers, who then took her back to her room. Later I learned that there was more to the story. The situation had to do with her not being able to depart with her latest boyfriend. I knew over the course of that winter she had gone through several boyfriends. More than likely, just by statistics, one or more of the boyfriends she had rejected would have been on the aircraft she departed on; thus, I did not see the problem. In witnessing the emotional meltdown in the hallway, I knew she was being ridiculous and that she would never receive such attention again in her life. Even with that, after giving it some thought, I learned to be careful when criticizing past winters as it hurt feelings. Every winter is special to most of the crewmembers, and if they were all alive at the end of winter and the station had not been destroyed, they were relatively successful. I wanted more than that for my crews.

"Toasty" is a term used at the South Pole to describe someone who is having mental issues during a winter. Scientifically it is known as Polar T_3 Syndrome. The causes of this syndrome have been debated, but whatever the cause, I did see a mental deterioration in some of my crewmembers over a winter. It usually consisted of forgetfulness, disheveled appearance, or just acting out of character. One winter, I saw a crewmember mentally deteriorate and go from pleasant and professional to surly and openly resentful. His boss did a masterful job of reacting kindly to the surly attitude and was always complimentary as to the fellow's technical abilities. A boss with a lower level of people skills would have had a hard time dealing with the situation and more than likely, in frustration, would have created a disaster. One day while walking the hallway, I greeted the toasty crewmember with a standard "how are you doing?" He responded with a very emphatic "fine, did someone say I wasn't?" He was not fine. Several crewmembers seeing his deterioration, took it upon themselves to help by eating dinner with him and watching movies together. I personally thanked a helpful crewmember for his assistance as I had noted him eating with the fellow at late-night meals. Knowing the toasty crewmember's unstable demeanor, I asked how he had managed to sit and eat with him. He said at times it had not been easy, and more than once he wanted to hit him in the head with the saltshaker. I do not think anything made me prouder of crewmembers than seeing them intervene, help in such volatile situations, and resist the urge to hit someone in the head with a saltshaker.

Others had problems as the long winters progressed that may have been indications to the validity of T_3 Syndrome. This syndrome is supposedly caused by decreases in the thyroid hormone T_3. The decrease in hormone can cause one to be forgetful and have mood swings. To counter the effects of T_3 Syndrome, many of us took vitamin D_3 as a supplement, which was said to help. Around midwinter, one fellow began making technical mistakes at his job. One of which was quite costly as he used a wrong tool and technique and destroyed an expensive part. He had a hard time focusing on tasks and once told me that he found himself walking down a hallway and did not know why he was there. In his case, he was such a self-aware and honest fellow that he recognized the malady and reported to the clinic for help. I understand he was administered medication. To his credit, he readily helped with other projects that were less mentally challenging and was genuinely liked and appreciated by most of the crew.

Throughout the three winters I spent at the South Pole, I made a practice of walking through the station and talking to crewmembers, preferably during their work. I had learned from many years of being in a leadership role at remote locations that it was important to be able to assess how crewmembers were feeling and doing. The best way to do this was through personal interaction. I made a point of personally greeting them when meeting in hallways and other locations. Using names is important, and fortunately for me, I learn and remember names easily. At the South Pole, especially during winter, crewmembers are used to constantly running into each other, and it is normal for repeated greetings. I once had a crewmember ask me if I really cared when I greeted them and asked how they were doing. It was an easy yes, and I explained to him that when I greet someone, I am assessing his or her overall condition. I had learned from experience that when one greeted a person who was usually upbeat and their response to the question "how are you doing?" was answered with a muted "okay," they were not okay. When I received such a lukewarm answer, I would then ask in more detail how the person was doing, usually by making a joke, and I would learn what was really going on and try to help.

While I tried to maintain my captain-of-the-ship persona, I did not want to be perceived as too rigid or uncaring, as the Ahab character had been. I did not worry about being popular or liked, but I did not want to be hated or appear aloof. I wanted to maintain a distance and to be taken seriously. It was better for the crew that way. While a gulf existed between us, it was important that they felt comfortable talking to me when they felt they needed to. Despite my self-imposed isolation, I always wanted to be available to talk and made a point that if anyone had a problem, it was my number one priority to help. Over my three winters, many crewmembers talked to me about problems they were having. The problems could be related to their winter or off-ice. Some were quite easy to solve, others more complex, and some were impossible. I found that many times it was not a solution the crewmember was seeking but just someone to listen. Early in my career, I was not as adept at this and would sometimes do other things while listening, such as continue to work on the computer. A few times I had said that I was multitasking. I now fought the impulse to do that and tried to focus my attention on the individual as best I could. During the most serious of these discussions, I would break character from the captain-of-the-ship persona and, if I thought it would help, reveal personal information about myself. Many of the problems the crewmembers were experiencing were things I had experienced in my wayward life. I could relate in one way or another to most anything that was happening to them, although it may have happened to me many years earlier. Sometimes we developed plans together and I offered specific advice. Sometimes all I did was listen. I always enjoyed talking with crewmembers and felt good seeing them feel better after a discussion. They probably never knew, and did not really need to know, that they and their problems were the most important things in my world.

I had knowledge of many past winters and knew some of the crews had little respect for their leaders. One WSM, who was highly intelligent, extensively utilized a voting process he had learned during teambuilding. I am sure he thought this was a democratic and fair way to make decisions. I agree it is, with things like what movie would be shown, but he was using it for more serious decisions.

Crewmembers from that year told me that while initially enjoying the process, after some time, they began to feel their leader should be making the more serious decisions. Some asked "if everything was a vote, why was he there?" I never had that problem and was always quite comfortable making decisions. After learning about the voting-run-amok situation, I recommended that teambuilding should not stress the voting process. Even doing that, I did solicit crew input, and when valid reasons for a change were presented, I would change my direction. I think it is stupid to hold fast to a bad decision based solely on one's ego. I cared very much for my crews and believed that every success was theirs and the few failures we had were mine.

There were two things I made extremely clear to my crews at the onset of winter. These things paid off for me and, I hope, for my crews. First was that their conduct and performance mattered. If they had any kind of desire to work foreign contract type jobs, doing a good job during the South Pole winter was quite important. The second was that I never went away. What I was saying by that was that on a South Pole winter crew most of the WSMs are "one and done." They do a single year and then go back to what they did prior to the South Pole winter. This is usually quite different than their job at the South Pole. The crew realizes that after the winter is over, they will more than likely never see them again. Sometimes crewmembers are happy with that fact. With my solid background in contracting at remote sites, I made it quite clear that if they did a good job, I would help them with other jobs if I could. I would also serve as a solid professional reference for them. I made it well-known that I was quite adept at doing that and could say things to their potential future employer that a standard HR person from the company they had worked for would not be allowed to say. Later, I was taken up on that offer many times with references for jobs, educational programs, and even housing referrals. My motive for ensuring the crews understood what I was willing to do was to instill in them early on that their performance and ability to work with their crew counted. This was necessitated by the fact that on some past winter crews, some of the crewmembers decided it really did not matter what they did because winter would end and there was no real penalty or reward for their actions. I wanted them to know that there was.

Another thing that I tried to promote among my crewmembers was the great honor it was to winter at the South Pole. Few people have ever done it, and it needed to be taken quite seriously. I knew of former crewmembers who had miserable winters and later, while off-ice, had an epiphany about how wonderful it had all been. By then, particularly if they had been a negative force on a crew, it was too late, and that is how they would always be remembered. I had that happen out on remote islands where I had previously lived and worked, and I had blocked a few former employees returns because of it. I considered my experience at the South Pole to be a great gift. I told my crews about the gift they had received and that it would be up to them how to accept it and what to do with it. I told them that THEY were NOT the gift. This was mostly meant for some of my younger crewmembers who had gone through life thinking that they were. Some embraced that and some probably ignored it, but my heart warmed when one of my younger crewmembers wanted to talk to me about it midway through a winter. She said she now understood what I was saying, she was not the gift, as she had at times been told in her young life, and now understood and appreciated the great gift she had received.

Although I did try to be accessible to my crews, I did not actively socialize, except for during movies. I supported all events and would stop by and say hello. I never drank with the crews until after we had completed our winter and the arriving summer crew was on station, completed their turnover ERT drill, and assumed the duties. I would then have a drink with them and wanted that to be a special event that meant something. While it was easy for me not to drink, I do remember a few occasions that were tempting. One was the making of Bloody Marys in the galley. I could smell them and wanted one so badly. Another was scotch tasting with crewmembers who really appreciated good scotch. In lieu of alcohol, I turned to the highly caffeinated sports drinks, which I developed a taste for. It was not the same, but it was something to look forward to.

My room had the advantage of being close to the B-1 lounge. The lounge would usually be quite active on Saturday nights, with much drinking, games, and conversation. I would walk through, say hello, and look for any signs of trouble. There never were any. Sometimes I would sit for a few minutes and join in conversations. At times I noted that my presence had an impact on the socialization as it would become slightly more muted. I wanted the crew to have fun. I did not want to intrude on their fun, so my visits were usually brief. They had earned their single night off to spend celebrating with their fellow crewmembers, and I was on duty.

While I liked and appreciated some crewmembers more than others, I made a point of never appearing to have favorites. I did have favorites. It would be difficult not to have favorites after witnessing some crewmembers trying so hard and others putting out less effort. The crews were filled with interesting people, many of them with extremely unusual histories. I could easily have made friends and been part of groups, but I made it a point not to. I knew that in past winters WSMs had eaten with some favored crewmembers at their own tables. I ate alone. I was outside alone, and as the winter passed, became more and more isolated. The advantage of this was that the crews never really knew me. Taking this to a bit of an extreme, I never played the music I would have preferred during my dishpit duties. I felt that my musical tastes were something personal to me that the crew did not need to know. It humanized me, and I did not want too much of that. I let the crews know what I thought they needed to know about me, such as work history, the fact that I was married, and that I had many cats. On occasion I would divulge more personal things if I thought they would help during a conversation with a crewmember seeking help or advice. I wanted to always be available to them if they needed something but not be overly sociable. Living like this helped me to remain an authority figure and not become compromised. I did not want to become just another crewmember. I was never that, and I thought it was best for them. I was so alone, but never lonely.

Although I resided in a world of mostly self-imposed isolation, I did have contacts with the outside world. Phone calls and emails to my wife were important events and kept me remembering there was a world beyond the ice. I would also get phone calls from my boss, Bill, in the Denver office, who honored me most by leaving me alone to do what I thought I needed to do. I enjoyed speaking to him as he understood what the crew was going through, and I knew he desperately wished he were there with us. I wanted him to feel the station was in good hands and that I would do whatever I could not to cause him more work, as he was very involved in long-term planning and other vital issues for the future. I knew the last thing he needed was a major problem or scandal at the South Pole that would necessitate his involvement, as had happened in the past. For the most part, I felt I accomplished that.

The South Pole in winter can be a harsh place for some. There is a small guidebook that a new South Pole winter crewmember will receive prior to the start of winter. It contains many great things and several pieces of advice from people who have previously wintered. One of the things that stood out to me was a statement to the effect that someone who is kind will get kinder and a mean-spirited person meaner. I saw that. I had a crewmember that I really liked to talk to as he had some fascinating interests and skills. He almost immediately had issues with the community, which initially started with his complaints about the food and some of the practices in the galley that he saw as unhygienic. That winter crew was incredibly tightknit and quickly took the galley's side, making the fellow a pariah of sorts. After he complained about the fact that kale grown in the greenhouse was being served too often, the galley staff retaliated and made sure kale was in most everything. I had never seen kale in pizza until that time. His reaction to the pariah status did not help, and there were several petty vengeful acts from both sides. One of the worst examples was someone putting fresh fruit in his baggage as we were departing the station at the end of winter. The hope by the miscreant who did this was that the fellow would get stopped by New Zealand entry authorities and get a fine for not divulging what he was carrying. Toward the end of that winter, when the crew was tired, there were several wild conspiracy theories, such as a sabotage of the greenhouse, where he was considered a primary suspect. Looking into it, I felt it just a late winter conspiracy in which he had played no part. As winter ended and time passed, it was viewed as ridiculous and was forgotten.

Another fellow that interviewed very well in Denver had issues about halfway through the winter. He was highly skilled, and by some time after midwinter, he had completed all the work that had been assigned him for the entire winter. We had been assured, prior to winter, by the full-time Denver person who authorized his hire that there would be plenty of his type of specialized work to keep him busy. There was not, and now, late in the long winter, I had a crewmember with little to do. This led to him doing other duties as assigned, which is never a good thing. His other duties as assigned consisted of deep cleaning the station, which he willingly did, but not with the same level of professionalism or enthusiasm as he exhibited while performing his specific craft. Then a side of his personality emerged that consisted of him posting political cartoons that were offensive to some. When questioned by his boss, he stated that he wanted to offend. The desire to offend is not a good trait for a South Pole winterover. He exhibited other antisocial traits and became a loner. Things escalated for the worse when the fellow received his performance evaluation a few weeks prior to the end of winter. He was incensed that he had scored low in community involvement. An interesting twist occurred when he thought I had done the evaluation. I had never even seen it. Soon after, a strange event occurred when a crewmember informed me that he had overheard the fellow having a long and heated argument with me while in his room. He wondered what I was doing in the fellow's room as my name was being shouted during the argumentative tirade. Oddly, I was not there. For three weeks prior to receiving our first flight, he walked around the station in a near catatonic state. He was on the first flight out. I think if he would have had enough work to keep him busy during the long winter, this could have been avoided.

Another younger crewmember started exhibiting bizarre behavior and hatred for his boss early in the winter. He became extremely withdrawn and edgy. Fortunately for him, his boss had previously wintered and understood its effects. His boss tried talking to him but seeing no improvement in their relationship, came to me for help. The crewmember in question was extremely skilled and was spending a lot of time working alone. Due to his working on night shifts, he would miss many crew activities. I talked to him on several occasions and remember an early talk during which I told him how valuable he was to the crew and what a great future he could have. He could have that if he successfully completed a South Pole winter. I was being totally honest as the fellow was a trained service representative for his craft with incredible skills. I would have given anything to have had such a skilled person out on the remote sites I had been to in the past. After his boss asked me for help, I walked down to the troubled crewmember's workspace, and we talked. We talked about the winter so far and how he felt about things. He was bored as there was not a lot to do after he had completed his work tasks. I asked him what he was interested in, and he said submarines. I told him when he had his work done to read about submarines. We then discussed the contractor world and the kind of job someone with his skills and qualifications could have in so many interesting places. I wanted him to know how valuable he was, *if* he could get along with people, and *if* he could successfully complete the winter contract. I then told him I was always available to help and would help him with future employment, but he needed to do a great job that winter. Soon after, his boss came to me and said he did not know what I told the fellow, but he was a changed man, and for the better. I was happy to hear that but knew there would need to be a few more such conversations prior to the end of winter.

I never slept well at Pole. In the summer, it is light twenty-four hours a day, but there are window blinds and other coverings where you can darken the room. In the winter, my window was covered with a shop-made wooden bookcase and my room was quite dark. Even with that, I never slept well, and there were different theories on why. One was the altitude, which seemed to change due to a physiological factor caused by changes in the barometric pressure. While the true altitude never changed, the human body would react as if it had as the barometric pressure raised or lowered. This factor meant it could go from its actual 9,300 feet to the body feeling it was at over 12,000 feet. This made breathing and sleeping more difficult. Another challenge for me was that when I would try to go to sleep, I was often deep in thought. This was especially the case during the third winter, during which we faced so many unknowns. This made relaxing and sleeping difficult. I learned to leave the deep thoughts and speculation and to take myself to a better place. I would transport myself mentally to an old bar in Fells Point, Baltimore that I loved or to a dinner with my wife in a lovely Rockport, Texas, restaurant. I took melatonin almost nightly, and that would get me through about four hours. After it wore off, I would get up, walk through the station, and if a satellite was available, check my email. I would go back to bed with mixed results. It was an awful experience, tossing and turning for hours, but many of us did that. I was in a slight state of exhaustion much of the time and had constant bags under my eyes. I always worried about my crewmembers when I saw them exhausted as that could lead to an accident or an unpleasant encounter with another crewmember over just a minor thing. Many of us were tired, especially by the end of a winter.

I tried to monitor myself for any signs of mental deterioration or unusual behavior, and I noted several things. One was increased irritation over minor station matters. Some of these things were so minor and would have been laughable in any environment other than a South Pole winter. Situations such as what I perceived as a negative tone in an email, an off-hand comment by a crewmember, or finding a minor mess in the station could be sources of irritation. Fortunately, I was able to see these things for what they were and could usually easily control my reaction to them. More difficult for me was dealing with things going on in the world office. At times I would rage at the national news that reinforced my feeling that our society was losing all the hardy self-reliance it had been founded on, that civility was out the window, and that we now lived in a silly mess. If I received a notification that my wife, Melissa, was struggling with something at home or that a cat was missing or was sick or injured, it would greatly bother me. These things I kept to myself. The main objective in such a situation was to do what I could to help from afar and make sure the crew never saw me in any kind of distress. When I had done all that I could, I found that getting outside and moving helped.

On a few occasions I experienced some short-term memory loss. I remember one event where I was having my morning fruit juice, which I would obtain from the galley. I usually put an ounce or so of energy/caffeine supplement in the juice, but on that day, I could not remember if I had added it. Not wanting to overdose on caffeine if I had, I did not put more in. Another time, while walking from my office to my room, I thought I was carrying my headlamp, but when I arrived at my room, I found I was actually carrying my office water bottle. I was deep in thought during the walk and hoped that that was the cause. I took the D_3 vitamins as a daily supplement and was never sure if they did any good, but I experienced no known harm from them. All in all, although at times physically exhausted, I made it through each winter and felt sharp enough to do what I needed to do.

A positive relationship with the Denver office is of vital importance during a South Pole winter. It is very easy for the relationship to turn negative, and I tried to not have that happen. The major challenge is preventing the very natural us-versus-them mindset from creeping in. This mindset can creep in because of things real or imagined that happen to people in isolation who are in contact with support staff or management from afar. The isolated people can quickly conclude that they are not being appreciated, are being ignored, and are being told to do things that, to them, do not make sense. Past wsms had different ideas about what type of relationship they wanted to have with the Denver office. Some used "Denver" as a scapegoat of sorts, blaming that office vocally to their crews for unpopular decisions they had to make. They would then tell their crews that they had been forced to make the unpopular decision by Denver directives. I found that ludicrous, as I had mostly free reign with my crews and received very few directives to do much of anything per a Denver command. The past wsms who did this were simply trying to dodge responsibility, and in doing that, they promoted a bad relationship with the Denver office and weakened their own position. I endeavored to show the Denver staff in the best light possible, for the good of the crew. The fact was that most of the Denver personnel were quite qualified and held technical information that we did not have, and many had previously wintered. Most of them understood what we were going through and wanted to help. Most of the Denver employees really did care about what we were going through and were available if we needed them. I remember a late-night call with a Denver-based technical expert when our crewmember apologized for waking him up. The Denver fellow quickly came back with a very genuine "please call me anytime you need to." We were all grateful and impressed. I tried to promote the idea that Denver was a very valued partner, and even a friend. This worked well my first two winters but was challenging during the third.

An incident that happened during my third and last winter had a long-lasting effect for me. One Saturday night, I was making my usual rounds through the station and came across a group of younger crewmembers in the TV lounge with the door shut. They were all laughing, drinking, and seemed to be having a good time. On the table was a deck of cards, and a game of some type was in progress. I said hello and one of them made a comment to me about joining them. This was met with a stony silence from the rest. It looked like they were simply having a fun time, and not wanting to impact that, I departed. A day or so later, I was informed that a game of strip poker had occurred on station, and I made a few informal inquiries into that. I was able to gather a few sketchy details, and it appeared to be a very minor issue compared to things that had occurred in years past. Even so, I wanted to find out a bit more about it and make sure it was not repeated. A day or so later I received a visit in my office from two crewmember who wanted my door closed. After the door closed, they went on to tell me that they had participated in the game. They stated that it had only gone to the point where a person or two were in underwear. They were quite concerned as they had learned that someone had written something to the Denver office regarding the event, and they thought they might be in trouble. To me the event as they described it was a trifle that would be taken care of on station. I was more concerned about how anyone outside the station even knew about it.

Several days later, the adage "what happens at Pole stays at Pole" was once again quite wrong. Several innocuous emails had been sent north by crewmembers who were not even present at the strip poker game. They were not really complaints but were sent simply as information to people at the Denver office. I did not see the emails, but I assumed the senders knew little of the actual event and that the facts were probably distorted and amplified. This is a quite common at the South Pole and is probably even an Antarctica phenomenon: something that allegedly happens on-ice, particularly a negative event, is amplified by an audience off-ice and begins to take on a life of its own. I was notified that human resource people in the United States were now involved, and that the event was being looked upon as some type of major act of immorality and breach of HR rules. Around this time there was a criminal act at the McMurdo Station, where a drunken idiot vandalized the chapel, and now, we were being lumped into that mayhem. I was requested to speak to an HR person who represented my employer, and I explained what I knew. I made the mistake of saying that the event was being overblown and so many worse things had occurred in the past. Much to her displeasure, I went on to state that I would not even report such a thing as we could take care of it on station and that it would never reoccur. This caused the robotic HR person to give me her opinion on what my responsibilities were. A person who had never been to the South Pole, let alone wintered, was telling me my duties and responsibilities. I was enraged but kept my cool, and at the end of the ridiculous diatribe, I merely stated, "Noted."

Sometime later, I learned that all but one of the companies and organizations that employed what I now called my "strippers" had ignored the incident. One company had not ignored it. This resulted in two of my crewmembers receiving formal letters in their employee files over the incident. I was incensed and so were the two who received the letters. After receiving the letters, they developed a unique defense. Their defense was the bad conduct they had witnessed during the summer by Denver-based employees. While I agreed with them, as I had seen such myself, it was not really a defense for their actions. The event should not have occurred. I went on to write my thoughts about summer conduct and a few other irregularities I had noted regarding Denver staff to company management. I received a response, but it seemed superficial and did not really address what I had written. I believe what I wrote may have cast me in a negative light for even mentioning the things I mentioned. I had witnessed a few similar incidents with Denver employees during the summers. During those events, I felt that rather than try to change what had gone on for years, all I wanted was for them to leave and for winter to begin.

By this time, there was a belief among some crewmembers that Denver-based staff who were being paid to work from home due to the virus had little to do. Having little to do, they were now dabbling in our affairs out of boredom. I shared the belief to a point but had experienced USAP working from home, and it had its challenges and was not as easy as one might think. I found while working from home that one's workday could be longer, as there was no defined cutoff point, and work then morphed into the entire day and evening. Still, the "stripper event" was not the only bit of dabbling we experienced that winter. I had noted some things that in the past were hardly looked at but that winter were getting more scrutiny. To exacerbate the situation, during this time my boss in the Denver office left the USAP and I no longer had his support or vast knowledge. With him gone, I had several weekly scheduled and at times irritating "tag-up" phone meetings, during which I basically said the same thing every time. I did share the opinion, up to a point, that there were some people off-ice with little to do, sitting at home getting paid, who wanted to appear to be doing something. I really did not want that "something" to be the conduct of my crew. It needs to be stated that I think the Denver full-time personnel, overall, did a great job supporting the winter crews. There were additional stresses faced by all during my third and final winter that were quite challenging. I will readily at least consider any criticisms sent my way and never forgot the great honor it was to be in my position.

There was a small positive that came from how the stripper event had been mishandled. The participants, particularly the two who had been singled out for disciplinary action, were not quiet about their displeasure. They knew that far worse had been done in the past and had received little to no disciplinary action. They rallied some of the crew in their support and seemed to focus their hatred on the HR person who had spoken to them and had informed me of my station duties. This served as a bond of sorts, and the crew knew the HR person's name. Soon after the event, during a movie, a troll-like creature appeared on the screen and a crewmember in our audience yelled out the HR person's name. Laughter ensued as other crewmembers understood the analogy and meaning. Still, the damage was done. Off-ice people had gotten involved with my crew and not in a positive way. I take full responsibility for the event happening. It should not have. But in the grand scheme of a South Pole winter, it was nothing. I was very protective of my crews, and this dabbling by people not at the Pole, who could never understand such events or how to deal with them, troubled me greatly.

CRISIS

During my first winter, an event occurred that was an example of a major problem at home one may face during the long isolation of a South Pole winter. It is also an example of how one can be affected by something happening a world away. Having a home on the Texas Gulf Coast, I accepted the risk of the annual threat of hurricanes. During hurricane season, I would watch the weather very closely. Because I was usually away somewhere else in the world, I had hired a fellow who owned a service that would automatically board up my house when a hurricane formed in the Gulf of Mexico. This would be after the hurricane's strength and direction were known and appeared to have a chance of striking my area. The one rule was that he and he alone would make the call to board up a house. This was necessary because people did not really want to pay for the process unless a hurricane strike was imminent. The problem was that he had so many homes to board up that he needed several days to accomplish all the work before a strike. The boarding consisted of the screwing in place of precut plywood sheets over all windows and doors. When properly boarded, a house was certainly not hurricane proof but would be much better able to keep its windows intact. This would serve to keep the home's interior and its contents intact and dry. My home was boarded a couple of times through the years, but as the hurricanes had not struck my town, the boards were quickly removed and stored and all went back to normal.

Seeing the weather report for my area, every day I tracked the progress of the tropical storm in the Gulf of Mexico and its growth into a genuine hurricane. My little town of Rockport, Texas, had not had a real hurricane in forty-seven years. It was a running joke how we would board up houses and then remove the boards, many times not even getting a sprinkle. While tracking storms, I would be looking for signs of two things. First, the storm changing direction and heading somewhere north or south of my city. Not that I ever felt good about it hitting somewhere else, but at least my worries were greatly lessened. Second, the storm losing energy and being downgraded in its potential severity, as so many had before. This storm did neither. I watched in excruciation over several days as the storm stayed on a direct course for my little Texas town. The day before it hit, a strike seemed imminent. Now it was only a question of the storm's strength. On August 25, 2017, at just about the time the station lost internet due to our satellite's position, the national news stated that Hurricane Harvey had come ashore at Rockport, Texas, as a category four hurricane. The hurricane had surprised many in the local community with its straight path and increase in strength, and due to that, my home had not been boarded up. It was a sickening feeling, and I could do nothing but head outside for a walk and imagine the worst. While walking and later in bed, I cursed God for the malady and more specifically, for killing my cats. I assumed massive water had come ashore and had graphic images of them struggling for their lives, along with so many other terrible things.

Melissa had evacuated to San Antonio with only two of our many cats. The rest would have to fend for themselves as best they could. Hurricane Harvey slammed the little Texas city, and soon, photos emerged of its devastation. Buildings I once knew were now piles of debris, and for two days, I did not know if my old house still stood. Excruciatingly, I waited for news. Finally, it came. The Baylor-Norvell home of Rockport, Texas, built in 1868, still stood. Slightly damaged, missing a window, many roof shingles, and the top of its original brick chimney, it stood. The initial photos showed it looking somewhat beaten up but intact, but standing proudly, looking out at Aransas Bay as it had for 150 years. After about four days, Melissa drove through the barricades that were placed to keep people out of the area and returned to our home. I will never forget the short video she sent of her entering the old house. She was crying, and so were many cats. While the old house suffered no serious damage, the place was a mess, with our trees and fences down, debris everywhere, and no electricity. Then the best news came, all our cats were alive! I hope God forgave me for my outburst, and I hope it was out of character for me, probably made worse by the intense isolation and my inability to do anything. What made this situation the most difficult for me was that the South Pole crew requires a steadfast leader. For the most part, the crew looks up to their leaders and watches every move and every nuance of their actions. In my opinion, at the South Pole, especially in winter, there is no time for a weak leader that can be overcome by off-ice events. Nothing overcame me to the point of the crew seeing or feeling it, not even the thought that my beautiful cats and home were gone. I was careful to show no outward emotion about the event and kept any real feelings about it to myself. Being in the position of WSM, the crew came first, over everything on- or off-ice. I was to relive this experience during my third year, to a lesser extent, when Hurricane Hanna struck south Texas. Hanna was a much smaller event than Harvey, but once again there were sleepless nights and worry, which I masked from my crew.

In January of 2020, I arrived at the South Pole Station at the start of my third winter slightly sick with something I had picked up in Christchurch or on the aircraft. Upon arriving at the South Pole it was apparent that many people were sick, the diagnosis being the annual "crud" that usually wreaked misery on the station's inhabitants. This year seemed worse, with more people sick, and I talked to one fellow who had been on my 2017 crew who stated that he had contracted several varieties of the illness and had been sick for over five weeks. He looked quite ill but had a sense of humor about it and was at least up and moving. My own symptoms were sneezing, coughing, and a general tired feeling. It was unpleasant and lasted longer than any illness I had ever had. I pitied the person on the other side of the soft wall of my berthing room. It must have been miserable listening to me sneeze, cough, and hack all through the night. Many people were suffering the same thing. Initially, we joked that we had contracted the newly discovered coronavirus, but our medical personnel informed us that the virus had not found its way to the South Pole. Sometime later, all joking concerning the virus stopped.

During mid-February of 2020, just days after the station officially closed, we were responsible for providing fuel and rooms for transiting Twin Otter and Basler aircraft on their transit home to Canada. Their flight path took them from wherever they were working in Antarctica through the South Pole Station and on to the British station at Rothera. From there they would transit the Drake Passage and continue their flight up the coast of South America. One day just prior to landing, we received a radio transmission from an arriving Basler that they were requesting transportation for the crew from the aircraft to the station. As the station is just a couple of hundred yards from the aircraft parking area, I thought the request a bit excessive as the crews always walked. My initial response was to say no, but I remembered the Basler pilot from an earlier transit and really liking the fellow, relented and radioed our heavy equipment operator to assist them. Later, I was glad I had changed my mind as the pilot arrived in awful shape. I met him at the entrance to the station at about the same spot where Buzz Aldrin had sat after arriving ill. The pilot was quite ill, with flu-like symptoms, and had to sit for half an hour before being helped up the stairs to the station's second floor. At the top of the stairs, he was so fatigued that he needed to sit and rest. We then used a wheelchair to get him to our clinic. He spent three days in the clinic being treated and on IVs. After he began to feel better, he was released to the main station, where he spent several more days recuperating. He finally improved to the point of being able to make the flight to Rothera, but as a passenger, not as a pilot. The concern of the station was that this particular Basler was known as the "Chinese" Basler. The aircraft and its crew had spent their season at Chinese Antarctic bases. We were informed by our medical staff that the pilot, while quite ill, did not have the coronavirus, due to his symptoms, but we had no way to test to make sure. As no one else came down with the malady, I assume they were right.

During my third winter, the COVID crisis dominated world news. People were dying, borders were closing, and new terms like "social distancing" were created. We watched from afar as toilet paper became an important item that now needed to be hoarded. We witnessed many people around the globe reacting with what appeared to us to be irrational and bizarre behavior. At the South Pole, we never needed to practice social distancing, did not wear masks, and were free to act as if COVID did not exist. During this time, we would receive messages or phone calls from people who would joke about how lucky we were to be where we were, away from the COVID. I remember a teleconference that included many USAP members back in the USA during which such a comment came up from one of the participants about how fortunate we were to be at the South Pole during this time. I quickly shot back with the fact that we had loved ones and people we cared for back there and absolutely no way to help. We were not dancing a jig at the South Pole and reveling in how fortunate we were. It dampened the meeting's mood, which I really did not plan to do, and I received several sympathy and apology emails afterward. For me personally, these times were hellish as I was worried constantly about my wife, cats (which I thought initially may have been able to catch the virus), and many others. Every day, when the internet was up, I would check the news, and it got worse and worse. I told very few that my little sister was at that time in the hospital, had contracted the malady along with another insidious condition, and was facing a limb amputation.

During March, the coronavirus was more commonly known as COVID-19 and was making international news. People were being infected and dying at an unprecedented rate. Maps showed the spread of the insidious virus from its alleged inception in China to its arrival at almost every region on the globe. It soon became apparent that the only region not affected was Antarctica. The USAP now faced quite a challenge, keeping the U.S. Antarctic stations virus free while receiving new people and sending others home. International borders closed and this exacerbated the challenge. The crew knew that the virus would affect our eventual release in November, but it was too early to foretell its full impact. We had a long winter ahead of us during which it would all be sorted out. We assumed that after a few months it would all be over.

At the South Pole nothing happened, no virus, no movement of people, and it got dark and cold. Initially reporters who wanted to know how we were handling the situation inundated us with requests for interviews. The National Science Foundation, NSF, took the lead in these matters, and a strict policy of no interviews from station personnel was initially in effect. This policy changed later, and a few interviews were granted. During that winter and with the virus, there was quite an upsurge in crewmembers wanting to talk to me. My door was always open, and I felt honored by their trust. That winter our crew suffered a family death, had several family members with the virus, experienced a couple of unforeseen divorces, and other serious off-ice matters. For the most part, I simply listened to them, and that seemed to help. I could do little for them but wanted to do all I could. A beautiful aspect of a strong South Pole winter crew is the fact that crewmembers will take care of each other. That winter the crew was exceptional in the kindness and concern that they showed to fellow crewmembers. It was very much needed, and I very much appreciated it.

At one point during the winter, when the virus was raging and there were many unknowns, I began to consider a worst-case scenario. In that scenario the virus had mutated into something more lethal that had overcome the world as we knew it. In such a scenario there would be no one coming for us at the end of winter. I was asked by several crewmembers what I thought about such a scenario, and I told them I was developing an unofficial contingency plan. One night in the galley, the subject came up between crewmembers who wanted to discuss ways we could get to McMurdo in such a disastrous situation. I then, for the first time, spoke of the rough plans I had formulated to date. First, we might not be going to McMurdo. If we were in a true end-of-the-world scenario, they might not really welcome us. I pointed out that McMurdo was close to New Zealand but nothing else and getting home from there in such a new world with limited transport might be problematic. I then divulged that in such an event we might consider another route and head in the direction of the Palmer Peninsula. The advantage being that we would be farther north with more temperate weather and a greater food supply and could survive there for some time. In addition, it was not so far from South America, although the dreaded Drake Passage would have been a terrific challenge. While this may sound like an adventure/science fiction movie plot, and I thought the chances it would need to be undertaken extremely remote, it was still considered and discussed.

That last South Pole winter was unprecedented with COVID ravaging the world. We continued to watch and wait for good news that the virus was waning or that a vaccine had been discovered. We watched and waited, and the news seemed to just get worse. We saw the deaths occurring worldwide increasing, and any thoughts that we were just facing a more serious flu season were dashed. Crewmembers were in contact with family members and thankfully, to my knowledge, no crewmember lost a loved one from the virus. Still, it created an air of uncertainty as to the future and how it would affect us and the ones we cared for. One day, while walking through the galley, I overheard a discussion concerning when we would be able to leave. We knew that the virus had impacted the ability to fly aircraft to get us out. Trying to lighten the mood,

I told them that we were in an *unprecedented* situation. I used that word a lot that winter, enough that it was printed on a square in Wayne bingo. I then stated that there would probably be a movie made about us. This lightened the mood as the crewmembers then began to very vocally announce who would play them in the upcoming movie. Two crewmembers wanted the same actor to play them. This caused them to try and determine which one of them it would be better for him to play. Other crewmembers jumped in with their ideas, to try and end the dispute. In the end, another actor was found that was thought a better fit for one of them. One crewmember didn't like the actor that another crewmember suggested was best suited to play him and was very vocal about that. I must admit, the actor suggested was a rather unattractive comedian that I would think few would want to say they looked like. One of the younger females, with absolutely no hesitation, picked one of the hottest Hollywood actresses imaginable. Several crewmembers made suggestions about who would play me. I was surprised by the characters they picked as they were certainly more handsome than I. They had fun doing that, and I accomplished what I had set out to do. Even so, it was only temporary, and I knew the crew was worried. We would be entering a new world quite different from what we had left.

I had learned years prior, while working out on remote islands and through two wars, that what constitutes a crisis could vary greatly person-to-person. I had learned that pets could be family members and always took quite seriously a crewmember that had a pet at home that was lost, sick, or had died. Fitness for duty under the extreme conditions of a South Pole winter was of paramount importance, and a crewmember upset over a pet issue might not sleep well or make the mistake of thinking alcohol could help. Several of my crewmembers had pets, and through the winters there were some issues. Fortunately, humorous stories concerning their pets were more the norm, and I enjoyed hearing them. My cats at home were the source of constant anxiety for me. With my travels, I had gone many years without pets, and the introduction of what ended up being a colony of cats, due to my kindhearted wife at home alone, affected me greatly. She accepted kittens, being told they would be later picked up and adopted out. They never were. What happened was a name given—Colton, Stevie, Ginger, Tina, Mary on and on—and a quick trip to the vet to give up any reproductive rights. That was the price of admission to the colony. I saw in them an innocence and a purity that I so much lacked. I watched kittens that had come from horrible situations struggle for their tiny lives. I had killed excessively in my life and had been humbled and changed by the experience with our cats. We had so many, and cats cannot really be controlled unless they reside solely inside, which ours did not.

My day at the South Pole would be off to a bad start if I learned of a sick or lost cat. A terrible case in point concerned a small black kitten that was part of a litter of five that had been pulled from the house of an old woman. They lived in absolute filth and squalor and had developed medical problems. Melissa took them in and did her best to help them survive. Three were near death, and one poor little fellow died one night sleeping on her chest. The other two were given "infusion" injections by a veterinarian that, hopefully, would increase their chances of survival. The veterinarian put their survival odds at fifty-fifty. They both survived and were named Mary and Tiny Elvis. Mary was blind in one eye and had other maladies. Tiny Elvis was black in color and had problems with his back legs. He earned his name due to some type of ointment that Melissa placed on his furry little head that made his hair look somewhat like young Elvis Presley's. I prayed to God they would live, and they did. One day a tragedy occurred when Melissa came home from work very ill. She immediately climbed into bed, not knowing that Tiny Elvis had snuck out of the house when she entered. He was larger than when he first arrived, but still a kitten, and he had not been outside before. Melissa slept many hours, and when she awoke, she noticed him missing. She was horrified and frantically searched for him, to no avail. She then posted ads, but he was never found. At the South Pole, I was enraged. I was enraged and had many terrible visions of what had possibly happened to the poor little fellow. I was enraged at what I saw as the weakness and sloppiness Melissa had exhibited by arriving home sick and not taking care to ensure he was safe. In my mind, there was no time to be weak when caring for other things, and I was initially very harsh when I spoke to her about it. I do believe my reaction and rage were exacerbated by the fact that I was at the South Pole in the middle of winter, so isolated and so powerless. My harsh comments to her were not needed as she felt plenty bad about the situation without them. Much later, a miracle occurred: Tiny Elvis was returned.

During my second winter, after being asked the question by multiple crewmembers, I researched the possibility of having a station dog but found it was forbidden by the Antarctic treaty. Dogs had been on the first couple of South Pole winter crews and were photographed as crewmembers. I understand the treaty but really think a station dog, or maybe even a station cat, would be a great thing for a winter crew. I wish we had one.

There were also many minor crises that a person would probably not face in normal society. One day Melissa sent me a photo of damage done by one of our cats to cheap vertical blinds in what we call our "yellow room." (Our old house's interior is painted in a variety of colors, and that room is yellow.) In the photo, the window blinds are broken and there are things that had been on display on top of a small bookshelf that are now on the floor. A minor mess, and I thought the likely culprit of the mess was Tina. Once, while I was home, Tina had distinguished herself by getting her head stuck in those same window blinds. My reaction to it at that time probably made matters worse. Initially, I thought nothing much of the photo, but while lying in bed that night it hit me! Among the things lying on the floor in the photo were poison darts from a blowgun that I had collected in the Amazon many years prior. The darts had extremely sharp tips dipped in curare, a powerful poison. I envisioned Melissa or one of our many cats stepping on them and being punctured and poisoned. I jumped out of bed and made a frantic call home on the Iridium phone to warn her. Melissa told me she knew what they were and had picked them all up. They were safe in a cabinet. I went back to bed, and then it hit me. I had just called from the South Pole to warn my wife that there were curare-tipped poison darts on the floor. Lying in bed, I pondered my life, how I ended up making such a call, and the fact that it seemed normal to me and, I assume, to Melissa. Another minor event, but a good example of how life in such isolation brings challenges, occurred when Melissa sent me a cryptic email asking if we had an axe somewhere at the house. The message was sent right before the internet went down. I sent back a quick reply asking her why in the world she would need an axe. She replied, "Don't worry I have it under control." I then pondered why she would need an axe, with my mind coming up with all kinds of horrific scenarios. I was relieved to learn later that she needed the axe to try to remove a small stump as she was planting things in the yard. In giving it some thought, I still did not understand how the axe would have worked.

THE END OF WINTER

Most South Pole winter crewmembers will say their winter ended quicker than they thought it would. During the first week of August, there is a slight glow on the horizon as the sun continues its rise. An interesting phenomenon occurs when, even though the sun is still below the horizon, light is refracted and reflected on the ice, causing a dim light. Although dim, there is enough light to begin to see outside, the dark outlines of buildings appearing first. The sun will reach the horizon at the very end of the month. It will continue to make steady progress above the horizon through September. The appearance of the sun is the first sign that winter is ending, but it is not really the end. The end is nearly three months away. It is important that the crew understands this fact as those three months can be a trying time. This is especially true for those suffering from the fatigue of a long, dark, cold winter. When the outside light is of enough intensity that the aurora cameras on the station's roof cannot function and are turned off, the window covers are removed from the station. Due to the heavy tinting on the windows, at first all one sees when looking out is their own reflection. This will change and as the sun continues to rise. The day will come when the windows become actual windows again. At that point, one can look out to the new icy world left by the winter storms. This is a busy time as many things need to be prepared to open the station for the incoming summer personnel. Interestingly, while the sun may be visible, initially there is no warmth. To me, during first light, being able to see the frozen station with a layer of winter frost, it always felt colder outside. There was also the first sight of the ice that had drifted around the back of the station. There, tiny ice crystals that had blown in from winter storms were in such numbers as to become minor mountains.

There is much excitement knowing the end of winter is close. The first major task in opening the station is getting the skiway back in order to receive incoming flights. After a winter and all the storms, the skiway bore no resemblance to what it was at the time the last aircraft had taken off. Now, the skiway was indiscernible from all the other ice and sastrugi that ranged for nearly one thousand miles or more in every direction. I was extremely fortunate during my three winters to have heavy equipment operators who were superb at what they did. These men were responsible for cleaning up after the winter storms, starting with the establishment of the skiway. They would do this by first cutting it out with a huge land plane pulled by a bulldozer. After that, a heavy drag was pulled, and the skiway was groomed to a surface that was smooth enough to land aircraft. They needed little outside assistance or advice. I would walk the skiway and marvel at the work that was done. Soon enough, I would be preparing airfield inspection forms that were necessary to officially open the skiway and receive incoming flights.

Next, the aircraft refueling system is placed back in position and in operation, after winter storms made a mess of the area where it stood in February. When it was removed at the onset of winter, I had photos taken and drawings and measurements done. This I was particularly adamant about my first winter, and to a lesser extent the next two. I needed to make sure we could, in our post winter fatigue, place everything back where it should be. Once the system was back in operation, the fuel was circulated and samples were taken. The fuel needed to be within specification before we could receive our first transit flight. South Pole fuel is extremely pure, with the low humidity and the lack of foreign objects that could cause contamination. The initial flights were transiting small aircraft consisting of Baslers and Twin Otters. They had made the long trip down via South America and were on their way to other Antarctic stations. It was always a strange experience meeting the crews of these aircraft. They were the first new people we had seen in eight months. One Basler crew brought us "freshies," which consisted of oranges, apples, pineapples, and avocados. I remember sitting in the galley with the crew when the galley staff had cut up the fruit so we could all have some. We sat and savored the things we had not eaten for months. The long table was quiet, and while the fresh fruit meant little to me, it meant a lot to the crew, and I savored just sitting there with them.

In days of old, polar ships returning from long expeditions were cleaned, painted, polished, and prepared for their entrance into a home harbor filled with excited family, friends, and the general public. I wanted to continue that tradition. We always cleaned the station from top to bottom. This was not a pleasant task, as people were tired and all that was on their minds was their imminent departure. Even though I was the stations leader, I found myself at times lethargic in the process and was happy that I had crewmembers who stepped up and spearheaded the effort. The goal of the cleaning process, which I explained to each crew, was that when the summer people arrived, they would enter our station and immediately see the pride we had for it and the care we had taken of it. It reflected on us and our winter. I had arrived at the end of another crew's winter and noted that it looked like all they had done was mop the floor. I wanted my crews to do better than that. In addition to station cleaning, on my crews, I promoted the idea that we would also do projects throughout the winter that made the station better than when we had arrived. Projects such as major reorganizations and improvements of the music, arts-and-crafts room, and library were just a few examples. I wanted the incoming summer people to note and give credit for what the crew had done. I felt a clean station with a few improvements were indications of a great winter and, more importantly, a great crew. I was always quite impressed with each crew's efforts with these projects, and crew pride was very apparent.

After receiving a few small transiting aircraft, the arrival of the first incoming flight containing USAP personnel is a major station event. There is much anticipation and preparation for that first flight, which will officially "open" the station. This flight historically brings in operations personnel who will operate the heavy equipment, work in the vehicle maintenance facility, and operate the aircraft fueling system. The flight usually arrives the last week or so in October. (Due to the impacts of the coronavirus, the end of my third winter did not have an opening flight until the third week of November.) The next few flights will bring in galley personnel and others. The galley personnel are especially welcome as the addition of summer stewards will end the winter dishpit rotation and take over the janitorial duties.

Many of the personnel on the first flight are veteran summer people known by most of the crew. Even so, to some, they arrive as strangers. I have heard comments from summer people about how they had departed in February as friends but received somewhat chilly receptions upon their return in late October/early November. Unless they have wintered themselves, they do not understand that they are now "summer people." The new winterover crew they left in February is now a tightly knit band of people who have experienced the long Antarctic night and all that comes with it. The arrival of the summer people and the new winterover people who will relieve the crew is, overall, an incredibly positive event for the outgoing crew. Some crewmembers may be a bit standoffish to the new arrivals, but they are still happy that the reliefs have arrived, and that they will soon depart. The new people and the flights bring good things with them, such as fresh food and vegetables. Most winter crewmembers have been craving such things, and a first banana, apple, or onion can be sublime. Most importantly, the flights bring the winterovers their well-earned release.

Interestingly, there can be a bit of a stigma to departing the station on the first flight. The first flight has a history of removing overly toasty crewmembers and the occasional discipline problems. It has thus earned the nickname "plane of shame." I was fortunate to never have to place a disciplinary problem on that aircraft. It was used for people trying to make commitments elsewhere, who had medical issues, or who were just no longer needed as their relief was in and had assumed the duties. There was a story about an electrician from a previous crew who was placed on the list of passengers for the first departing aircraft due to convenience and space availability. He had no negative winter issues. When informed he was going out on the first aircraft, he became quite emotional and tried to stop the process and asked to leave on a later flight. He felt that others, noting his departure on the first flight, might see it as an indication that he had done something wrong or was having mental issues. No matter what aircraft crewmembers departed on, there were always very emotional goodbyes. Tears would sometimes flow in the hallway that led to the station's main front door, "Designation Alpha, DA," where outgoing passengers waited for the radio announcement that the aircraft was ready to load passengers. By the end of winter, incredibly strong bonds between winterover crewmembers had been formed that would never be broken by mere distance.

More flights arrive, bringing in new people, and with that, the station dynamics begin to change. During the first few arriving flights, the winterover crew still outnumbers the arrivers, and it is still their station. There comes a tipping point when the station is no longer "theirs," as winterovers depart and more and more summer people arrive. For most of the winterovers, it is a great relief to see the new arrivals, as they know they will soon be leaving. Sometimes there is a slight resentment. I remember watching a TV series that our crew started near the end of winter and continued into the station opening period. It was a Sunday, and I was sitting in the B-3 lounge when several summer people came in. I did not understand why they wanted to sit in on a series that we had been watching and was near its ending. Newly arrived summer people sat on either side of me, and I found myself silently resenting their presence. It was strange as I knew these people and I knew my feelings were irrational, but I did not want them in there with us. They were in our chairs, in our event, and they were not us. They would never be us.

During the turnover period, the incoming crew begins to assume all the station duties of the outgoing crew. One of the interesting things about the relationship between summer people and winterovers is a blame game of sorts. Many winterover crewmembers get the idea that they have done things over the winter that the summer people should have done prior. They at times also think they are the first ones ever to perform certain projects and ponder how they went undone for so long. Sometimes that may be true, but for the most part, probably not. The summer people on the other hand sometimes blame the winterovers for not doing things or not doing them right. Summer people are notorious for blaming the winter crews for mistakes and shoddy work and are known to exclaim phrases like "what the hell did they do all winter?" Most departments do this to some extent, but the vehicle maintenance facility, VMF, takes it to a higher level. The first group of arriving summer people are operations personnel who immediately begin to utilize the station's heavy equipment. That being the case, the work that was done on that equipment over the winter will be closely scrutinized and sometimes criticized. For one of my winters, the VMF shop foreman, knowing this would happen, took extensive photos and kept detailed notes of all the work that had been completed to prepare in advance for any negative comments on what had been accomplished that winter. I did my best to try and minimize the negativity, but it is nearly a South Pole tradition and difficult to control.

During the station turnover, there can be, on some crews, a hard line driven between the outgoing winterovers and the new arrivals. That can be best seen by seating in the galley. For some winter crews the seating can be quite segregated. The departing winter crew would have its own table with few, if any, new arrivals seated there. They sit and tell stories of their winter or talk about upcoming travel plans. During my first arrival, I watched the outgoing winter crew seated tightly together. They were quite noisy at mealtimes, and it struck me that they wanted to announce their presence to all interlopers by their sheer noise volume. Every day the group got smaller as crewmembers departed, but they stayed noisy until the end. Then they were gone. The galley was quieter, but as all winterovers know, they became immortal, as what they did during their winter will live forever.

There is much socialization between the outgoing winterover crewmembers, especially after the ERT turnover drill is completed. At that point, the entire crew can enjoy a drink together. I enjoyed these events immensely and would have a drink with the crew. The end of my first winter stands out the most. I remember sitting in a chair in the gym as a bottle of scotch was passed around. I took my turns and sitting there with those people, never wanted it to end. The next two winters' endings and socializations were similar, but not as intense for me as that first experience. At that point, past differences between most crewmembers had evaporated, and we were now joined forever. Summer people and a few of the newly arrived winterovers would sometimes join in these events and were accepted as guests. Unless they were an attractive female, for the most part, the outgoing crew ignored them. They were not us.

One of my great South Pole winter highlights and most cherished memories at the end of winter was the Antarctic Service Medal ceremony that occurred when my boss, Bill, arrived from Denver to relieve me and get the summer started. He would bring our medals and certificates, and the crew would then assemble in the galley. I had witnessed a horrific ceremony during a year when the wsm had sat and said nothing. He had experienced a lonely year and had become alienated from many of his crewmembers. What should have been a testimonial to the great work his crew had done was not. I vowed never to have such a ceremony and never did. The ceremony was always fun and euphoric as the crew would revel in their comradery and accomplishments and many inside stories were told. The medals were distributed and many photos taken. The winter crew had now joined the exclusive ranks of approximately 1,500 people who had spent a winter at the South Pole. The medal ceremony, to me, was the event that truly marked that winter was over. Even so, the crew would live on.

I very well remember attending the outgoing crew's medal ceremony at the start of my first winter and holding them all in awe. They had done it, and I was just starting. I specifically remember one person receiving the silver winterover clasp that signified the attainment of three winters. I was in awe of the significance of the medal itself, but the silver clasp for three winters was utterly amazing. At that early point, I began to think of the possibility of achieving that goal. I received my own silver clasp in a little envelope days after we had held a private medal ceremony prior to station opening. This was done because of covid and the fact that upon the arrival of the summer people who were bringing us the medals, the station could have been in a virus-condition, yellow status. This would have entailed masks and social distancing, and it made a normal medal ceremony impossible. Knowing this, we conducted the ceremony without having the actual medals. Now I see the little silver clasp itself as insignificant, but the value of the experience to attain it was priceless.

The day a winterover departs from the South Pole is an amazing experience, especially if it is a first winter. With the challenges that Antarctic weather provides, there may be multiple attempts to leave and disappointing aircraft cancellations. There is also a tradition to not remove one's bedding until the aircraft is in the air and close enough to the station that it will more than likely land.

This tradition can be taken quite seriously. When an aircraft is in flight but cancels, for whatever reason, and returns to McMurdo, the people departing seek the crewmember who removed their bedding too early. Even with delays for weather or mechanical issues, the day finally comes when an aircraft is inbound, and it is going to land. The departing crewmembers meet at DA, where well-wishers see them off. The atmosphere is usually chaotic and quite euphoric. The departing winterovers are easy to spot, with their long hair, winter beards, pale complexions, and sometimes fatigued demeanor. A radio announcement informs the depart- ing winterovers to walk out to the aircraft, where they stand in a group and await the Air Force crew chief's signal to begin boarding. When the signal is given, there are last goodbyes, tears, and cheers as they walk single file toward the awaiting aircraft. There is a tradi- tion among those who want to someday return to make sure they take a last look back at the station while outside on their way to the aircraft. It is a fact that some people departing the station after a winter believe they will return, but things then come up, such as a not physically qualified, NPQ, or a situation at home that ends up preventing their return. For my first two winters, while walking toward the aircraft, I looked back at the station and hoped things would work out with my PQ process, that other matters would not arise, and that I would thus return. I was fortunate and did.

For my third and last departure, I had mentally prepared myself to not look back. I had accomplished three great winters, had the Antarctic Service Medal with silver winterover clasp, and was leav- ing on an extremely positive note. There was really nothing else for me to accomplish, and I had long ago learned to leave remote assign- ments that I loved before they became just a lifestyle and a com- fortable routine. Prior to my departure, I had given some thought to that last walk to the aircraft and decided I would not look back. Walking with some concentration and looking only forward, I did not look back. We were leaving on a Basler, as there were no LC-130s due to the virus. I boarded the Basler and took the last seat on the left-hand side in the back. Looking up to the front of the aircraft, I was stunned to see the station with flag flying centered in one of the windows. It was like a photograph. From the air, I looked out the window and saw the station and the airfield distance markers, one mile, one and a half miles, two miles, three miles. Those very markers that I had struggled to find in extreme darkness and bitter cold. I had stood next to them under unreal conditions, and now, from the air, they were just tiny black rectangles on an icy surface. From the air they looked so close together, but I knew the true distance, measured in icy steps in the dark.

Upon boarding the aircraft there is a great sense of relief, as it is now quite clear you are leaving the South Pole, and for many, the knowing that they would never return. The flight to McMurdo on an LC-130 is about three hours long. It is slightly longer for the Basler, which I left on at the end of my last winter. Those flights for me were very emotional, with a mix of thoughts on what I had done over the last year and what was ahead of me. Upon landing at McMurdo, you emerge into a new world, quite different than the flat icecap you have just come from. It is warmer and mountainous. Immediately you see new people who you have never seen before and make a somewhat shocking drive to the station. For most, it is the first time they have been in a vehicle for over a year. Upon arrival at the station, you enter a world with streets that are not simply ice; they are dark in color and volcanic in origin. There are many buildings with unknown people moving from one to the other. Vehicles are everywhere. You enter the back of the 155 building and are issued a room key. You make your way to an unfamiliar space in a world of strangers. You will go to the galley at the first opportunity, taste food you have not had in months, and see more people than you have been around for a year.

The first time I made the trip to McMurdo at the end of the winter, upon arrival, I was requested to visit to the "Chalet" to see the NSF manager. There I was involved in a discussion designed to minimize any kind of potential problems with my newly released and very exuberant South Pole winterovers. South Pole winterovers on their way home have a history of going on drunken sprees and causing minor mayhem soon after arriving at McMurdo. Some of the McMurdo folks were not pleased to see us as we had a reputation for aloofness and trouble. I had heard the term "stinking Polies" used to describe us. I think this term may have come from the fact that we were known to only take two, two-minute showers a week. I am sure that some South Pole winterovers probably did not smell all that fragrant upon arrival. It may also have been just another term of endearment. At McMurdo, we were free to do whatever we wanted, the main requirement being no mayhem. I was requested to ensure everyone would be fit to depart—meaning not intoxicated at flight time. I have not heard of any departing South Pole winterovers ever missing that flight, although I know there was much intoxication while there.

The flight from McMurdo to Christchurch was always somewhat surreal. At that point, my responsibilities to the USAP and my crew were completed. I was just Wayne White on my way home. That transition was always a strange experience, but I did enjoy the relief of it. My first two departures were on U.S. Air Force C-17s, huge jet engine aircraft with plenty of room and an actual toilet. The quick flight and landing in Christchurch with the rapid change from the frigid to the subtropical was a wonderful sensation. Exiting the aircraft wearing the issued USAP ECW, we proceeded through a customs and immigration check and then through a section of the airport. We then walked to the USAP CDC where we turned in our issued cold weather gear. It was not a formal process and simply consisted of throwing the stuff that had protected us for a year onto piles. It felt good to be free of the gear. Outside, vans were waiting with a small luggage trailer to take us to our various hotels. Everything was so green! No cold and ice. Now grass, trees, kids, dogs, cats, and many other things we had not seen in a year. Whatever hotel we stayed at, the rooms were palatial compared to what we had come from. We had everything we wanted at our fingertips, but there was something missing, our crew.

Walking around Christchurch, I would sometimes run into crew-members, and it was always somewhat odd to see them in such a different setting. I was always very drawn to them, as we had experienced so much together. At the end of my second year, I received a message that several of our crew that were in Christchurch were going to have dinner together at a beautiful restaurant. I walked to the restaurant and there they were, over a third of the entire crew! I sat at the end of the table and talked. It was an overall happy feeling to be there among my crewmembers, and yet I was slightly uncomfortable. These were people who I had shared a unique experience with in an unforgiving cold and dark place, and now we were back in civilization. It did not seem right, and I must confess to feeling slightly uncomfortable and out of place. I ate, drank, and talked with them and was happy to be a part of it. But after dinner and some talking with crewmember's, I made a hurried exit and went back to my hotel, where I was once again alone.

My last trip home was memorable. Aboard the Basler from the South Pole Station, we were required to wear masks, which was still not normal for us. At the end of winter, we had practiced wearing them. At a prearranged time right before our departure, the station went to "Condition Yellow" and required their wearing. After takeoff, we removed them. The views from the aircraft were stunning as we had opened the station so late that year that the temperature had significantly warmed. This kept the aircraft windows from freezing up. Traveling down the Beardmore Glacier and seeing vast crevasse fields, I marveled at Captain Scott and Ernest Shackleton's great achievement of traveling up and the down that space. Arriving at McMurdo was like previous years, but this time, I must admit, I did not have the same level of exuberance I had felt at the end of my two other winters. McMurdo was in "Condition Yellow" and masks and social distancing were required. The station seemed deserted as so few people were there due to reduced staffing requirements caused by the virus. Eating in the galley was odd as people were mandatorily spaced so far apart. This created a scene quite unlike that of a departing South Pole winterover crew sitting packed at a table, which would have been the norm. I was always impressed by the fact that many South Pole winterover crewmembers desire the close company of as many of their fellow winterovers as possible while in transit home. That was not allowed to happen in the McMurdo galley, but at the barracks, where we were all billeted, crewmembers gathered in a lounge in a circle, with masks on and off; relived South Pole memories; and discussed the future in a new world.

During my last winter at the South Pole, I developed a new appreciation for McMurdo, after witnessing the dedication and great work ethic of a crewmember who was a long-time McMurdo worker. We had several discussions on McMurdo, and he once told me what McMurdo people thought of Polies. He told me that Polies who had been sent to McMurdo for training or to assist during the summer were viewed as aloof and lazy. It was felt that the work pace at Pole was not anywhere near the McMurdo pace. As he was truly one of the hardest working people on my winter crew, I listened. In addition, I had seen photos of spectacular sunsets and the beautiful mountains around McMurdo Station and knew its rich history with Captain Scott, Ernest Shackleton, and other early Antarctic heroes. They had spent much more time there than their few days at the South Pole. McMurdo has a beauty all its own, which I had not been receptive to previously. I had been captivated by a desolate icecap.

Due to changes caused by the virus, there were no U.S. Air Force C-17s on-ice my last year. Several C-130s were in service, which made the flight from McMurdo to Christchurch longer. It also added more of an element of unreliability to the actual day we would be flying, due to maintenance issues with the aging aircraft. There were several weather cancellations, a crack that occurred in the ice on the McMurdo skiway, and a flight that turned around and returned to

Christchurch with a mechanical issue. The day finally came that
we departed. It was a New Zealand C-130, and we were packed in
like sardines on the nylon webbed seats facing each other, with little leg room. No one complained as we were going home.

We arrived in Christchurch extremely late, and even though no
masks were required in New Zealand, for some reason, we were
told to wear them on the shuttle ride. We were dropped off at various hotels, and the masks came off. Instead of the normal downtown hotels we usually stayed at, we were taken to motor lodges
some distance away. They were paid for by the USAP and were modern, clean, totally adequate, and far nicer than our South Pole or
McMurdo rooms. Even with that, after one night there, I checked
out, took a taxi, and checked into the Heritage Hotel downtown,
all at my own expense. I received a notification from the local USAP
travel people that they were not happy with my checking out of the
hotel that they had provided, and for the most part, I ignored it.
The mistake I had made was telling the USAP-paid-for hotel that I
was checking out—which I did to save the program the money for
the room. I had never, in the past, filled out an expense report for
my transit through Christchurch, and I did not that time either. I
was now totally on my own and wanted it that way.

While at the South Pole, we had heard stories of nearly empty
international flights. This must have been the norm at the onset of
the virus crisis. By the time we returned home, flying from Christchurch to Los Angeles, the airlines had canceled most of the regular flights and were flying the reduced number quite full. Mine
was. I had prepared myself for a miserable twelve-hour layover in
Los Angeles, but it turned out to be almost pleasant, with easy
check-in aided by extremely kind and professional airline repre-
sentatives. LAX seemed nearly deserted, and I had my first experience with the economic impact of the virus when I saw so many
closed shops. The airport should have been teeming with passengers and the shops doing a brisk trade. Not now. I spent my twelve

hours in relative luxury, with internet that continuously worked and great choices of food and drink.

I arrived in Corpus Christi, Texas, on a Friday, and Melissa was at the airport waiting for me. We were both wearing masks, and even after a year apart, there was no hugging or welcome-home kiss. We had no physical contact. She drove me to our home in Rockport, where I spent ten days in self-quarantine in our guest house. I had concerns, even though Antarctica and, for the most part, New Zealand were thought to be virus free. I had been on long flights with long layovers in airports and may have contracted the virus. This was of great concern as she is a teacher, and while I had no real fear of the virus for myself, the thought of being the cause of her or her students being infected greatly bothered me. Being alone in the guest house did not feel at all unusual and had one fantastic perk, I had cats, lots of cats!

DUTY, HONOR, AND GLORY

Being back in "civilization" after a long absence was something I had grown very used to. Even with the virus, small-town Texas was almost as I had left it. The differences were that masks were required when visiting businesses and I had little social interaction with friends. There were no invites to our old home, and visits with local friends were now talks from a distance and waves. I began to see that the intensity of the three South Pole winters, particularly the last, back-to-back one, had certainly affected me. I remember a sunny walk in our local beach park where Melissa was telling me something about one of her students. She did not know it, but I was not there. Upon entrance to the park, I had heard the park's flags waving in the brisk breeze. I was instantly back to the frigid, windy darkness, hearing the snap of the flags on the flag line and feeling and hearing the crunch of the ice, the glorious ice, beneath my feet. I had been alive with my three crews in an almost alien world, a world of isolation and cold. Cold—beautiful cold—with duty, honor, and glory.

It would be impossible for me to answer, if asked, which crew I considered the favorite of my three. I could not say I had a favorite, but I could state that each crew had different attributes. My first year's crew was my first year's crew and that will always have a certain magic to it. With my second year's crew, I had the most input into their selection. For my last year's crew, I had the least involvement in their selection and by far my most challenging winter.

I was involved in most of the selection of my first crew and during the interviews, used the experience I had gained in my years of working at remote sites. My first crew did not contain many veteran winterovers. It did have a high level of maturity and was not a young crew. It was my first winter; I used what had worked for me previously at other remote sites, and it worked. That crew had an additional four people who were assigned construction-type duties. Because of that, many extra tasks were completed that winter, such as tank leveling in the fuel arch, a flooring project in the main station, and several others. There were no serious personnel problems, and that crew went on to raise the bar, as far as South Pole winter crews were concerned. I won a company leadership award that winter, and while thankful for it, I found it slightly uncomfortable. It was really the crew's award.

For my second crew, I was very actively involved in most of the crew selection and used knowledge I had gained during my first winter. With that experience and knowledge, I was more active while on the selection panel and was able to tactfully override a couple of what I thought would be potentially problematic crewmembers. I actively recruited among former South Pole winterovers and in the end, was able to put together one of the most veteran crews that has existed at the South Pole. The crew consisted of almost 50 percent veterans of previous South Pole winters. That was an unusually high number. The crew was around 70 percent veteran, when you included those who had USAP experience, either summers at the South Pole or McMurdo summers and winters. The qualifications the crewmembers held for most positions were higher than normal, and that crew was capable of achieving superstar status. They metaphorically "ate the dogs." However, no crew is perfect. There were a couple of individuals who were not well-liked, and the veteran crew showed little mercy on them.

As I did a back-to-back winter, I was not involved in the selection process for most of my final crew. There was an assurance that I would get good people, and for the most part, that was true. What

I got was a much younger crew with a strong core group of millennials. During the winter, I was able to observe and live with the groups' strengths and weaknesses. They were sometimes silly and immature, but they had great command of technology and were extremely clever. They were also very kind to each other, and in the end, that mattered as we faced more serious off-ice issues that winter than my previous two combined.

Interestingly, I now know that if I had the possibility of selecting a "dream crew," where I could choose anyone from my three winters and put them all together, the results would probably not be anything close to perfect. While the person I might choose for a specific position may have been a favorite who was highly qualified and had a superb previous winter, there is the unknown factor of how that person would interact with new people. I think there are many ways that a WSM can successfully lead through a South Pole winter. Some in the past have been staunch authoritarians and have had miserable winters. Some did truly little leading at all. An end-of-season report by a past WSM stated that the WSM position was not really needed in a full-time capacity and could be done as a collateral duty. He had a good crew, faced no serious issues, and made it through a successful winter. There are different ways to do it, and mine was simply mine.

Other leaders, in the past, had been closer to their crews, as far as socialization was concerned. This sometimes worked out and sometimes did not, depending on the makeup of the crews. Although I have certainly done it, I think it amateurish to criticize past South Pole winter leaders. I was by no means the perfect leader, but I used the leadership style I had developed and always tried to make sure my crewmembers and what they needed came first. The crew must come first. To be an effective South Pole leader is not an easy job, and most that do it never do it again. All South Pole winter crews to date can be considered "successful" if success means the crew is all alive and the station is still standing and operational when the summer people arrive. I wanted more than that.

I wanted to lead crews that understood and appreciated those that had gone before them, the great heroes of the past. I wanted my crews to be proud of what they had done. I wanted them to genuinely care about their fellow crewmembers and leave that place the proud recipient of a well-earned gift. A great gift that they had earned, understood, and would cherish forever. I wanted to command my ship with the power of the great Ahab, but an Ahab who had turned from the darkness and had seen what was most important, his crew.

One night at the end of my last winter, I lay in bed exhausted after lack of sleep and a tough outside walk. I pondered the effect of the back-to-back winter, it being my third. I realized I would be going home, more than likely, never to return. It was an emotional moment, and I realized I did not want the winter to end. While I needed to get home to my lovely wife and cat colony, the South Pole in winter was where I thought I belonged most. I thought back to my crewmembers from the various years, particularly the younger ones, and became quite emotional. I remembered their names, faces, and many interactions I had with them. In my life, I had never had children, but there in that frigid darkness, I almost did. It was overpowering, and I actually teared up. I thought of the poignant ending of the novel *Goodbye Mr. Chips*, and I did the same: Barricklow, Chen, Dudley, Eberhardt, Hall, Hampton, Precup, and so many more.

Roald Amundsen, Captain Scott, and Ernest Shackleton are now long dead. They are forever entombed at the sites of their great triumphs. Two reside in the ice and sea and one on a rocky island. There were no ordinary graves for these great men. They were the giants who faced great adversity with little margin for error, and their safe return was never assured. The men who built and staffed the original South Pole station are getting older, and each year there are fewer of them left alive. What they did to bring the South Pole station and its current scientific mission into the modern world should never be forgotten. While I cannot claim any kind of Antarctic discoveries, I can say that I did my best while at the South Pole and will continue to do my best to keep the memory of these great men and their magnificent deeds alive. We all walk in their shadows.

After my return from the icy world, I have had many people ask me very vaguely worded questions like "so how was it?" When asked that question, my mind races as I could think of a thousand things that happened while on that icecap. The darkness, the cold, and the warmth of my crews. So many things that so few people could possibly understand unless they were there and experienced it themselves. Sometimes, after a short pause, I give them the only answer I know that seems to fit, "It was cold." After experiencing cold at a level below what most humans can imagine, I developed a love for it. Not simply chilly-sweater weather, but lethal, subzero cold that forced me to learn to dress in a protective manner and understand its dangers. I loved the feeling of being alone outside in motion, with only layers of fabric, hide, and fur protecting me from an icy death. I loved being outside in that frigid darkness, my world now reduced to what I could see with my headlamp, moonlight, or the light from the spectacular auroras. It was there that I began to understand that I was never truly alone, as the great force and spirit that created the universe was always there. Standing out there alone, I was absolutely enveloped by it. The ice and the cold gave me that.

No one ever asked me what I did during the more than four thousand miles I walked. If they had asked, I would have told them that I spent the miles thinking. Thinking about so many things was the majority of what I did out there. Thinking about where I had been in life and where I still needed to go with the time I had left. Hours and miles spent thinking about my wife, cats, and things they needed. Thinking about issues with my crews and trying to develop solutions. Issues that would have been minor anywhere else in the world but in the winter at the South Pole loomed so large. While at the South Pole, I would never tell a soul the whole story, but sometimes, some distance from the station, on the darkest and coldest of Antarctic nights, I did something that for me was quite unusual. On those dark, frigid nights under the magnificent light of the auroras, I was as alone as most anyone can ever be. Almost alone, but with one magnificent and benevolent witness. On those nights, I was able to do what for me, rigidly riding those iron rails through life, was quite unthinkable—I danced—and sometimes with my wife, who was so very far away.